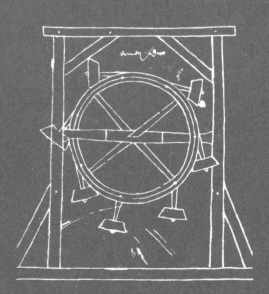

GREAT MEN OF SCIENCE

PURNELL

Contents

Left panel, top: Analysis on a special scanning table of the tracks of atomic particles taken by a heavy liquid bubble chamber. Centre: One of the Apollo 12 crew sets up an experimental package on the Moon's surface. Bottom: A celestial sphere.

Editorial

Authors
Edward Holmes
Christopher Maynard

Editor
Jennifer L. Justice

Illustrations
Oliver Frey

Cover: Michael Faraday demonstrates the principles of electricity at the Royal Institution.

Title page: An analytical balance of 1870. The balance is an important laboratory tool for chemists.

Published by Purnell Books,
Berkshire House, Queen Street, Maidenhead.

Designed and Produced by Grisewood & Dempsey Ltd.
Paulton House, 8 Shepherdess Walk, London, N.1.

SBN 361 03270 6

© **Grisewood & Dempsey Ltd. 1975**

Printed and Bound by Interlitho, Milan, Italy

Great Men of Science

Science, says the dictionary, is knowledge of general truths and universal laws gained by observation, testing, and recording. And the story of science is the story of men with enquiring minds who, since prehistoric times, have sought to explain the world about them and the sky above them.

The great men of science are those who combined the work of their predecessors and their own observations with such genius that they arrived at theories which revolutionized Man's understanding of his world. Men like Copernicus, Newton, and Darwin overturned prevailing ideas about Man and the universe, and in doing so, invited attack from established authorities. Other men, like Carolus Linnaeus, spent their lives ordering and classifying existing knowledge, sweeping away many of the misconceptions of the past and revealing new and significant patterns in nature. Their achievements were perhaps less dramatic, but were vital to the progression of knowledge.

Many of the great scientists were men of imagination and foresight who applied their findings to solving practical problems. For the line between pure science – researching and recording for its own sake – and technology, the application of that research, is often a thin one.

Great Men of Science traces the story of some of these men and their work, from the investigations into astronomy and medicine of the ancient world to the research into subatomic particles and the mysteries of the double helix of DNA.

Above: An early microscope – a tool which opened up whole new areas of research for scientists. Below: 'Experiment with a Bird in the Air Pump', by Joseph Wright (1734–1797). Such experiments, fashionable in the 18th century, were among the earliest investigations into the gases that make up the air.

The Story of Science

The history of science begins with Man's first attempts to find a constant pattern in the world around him.

The earliest men had no time for science. Their wandering and hunting way of life was a grim struggle for survival, and they had few opportunities to ponder on the reasons for what happened around them. But then Man changed from a hunter to a farmer, and the story of science begins.

Not surprisingly the first sciences to develop were those concerned with everyday life in the early civilizations. These were agricultural communities, so the timing of the seasons was immensely important. Men learnt to tell the time of year by the position of the stars – and from their observations grew the science of astronomy. Another very early and practical science was geometry, used in building and marking out land.

As civilizations developed and life became easier, the early scientists had time to speculate about the workings of the universe. At first they were inclined to explain what they could not account for as the whim of a god. But the early Greek thinkers tried to find more rational explanations. Their attitude was scientific: they believed in

observing and noting facts, and undertaking organized experiments to discover the nature of matter and the character of the universe. They built on the knowledge of earlier peoples, but believed that such knowledge should be checked by what they could see for themselves. The Greeks combined abstract speculation with practical common sense: Democritus' astonishingly modern theory of the atom was counterbalanced by Archimedes' orderly setting out of mechanical principles.

The Greek schools of science still flourished under the Roman Empire. But when the Empire crumbled under the onslaughts of the barbarians, men had little time for meticulous observations or for theorizing. Most of their books were destroyed. Although in the East the development of science continued unchecked, in Europe it was eclipsed with the coming of the Dark Ages. But the Arabs preserved the Greek learning and built on it; and through translations of Arab works science returned to Europe from about the 10th century.

The people of the Middle Ages believed in authority. Their views on religion they took

In prehistoric times men made life-like paintings of animals on the walls of caves. These may have played a part in magic rituals designed to help hunters.

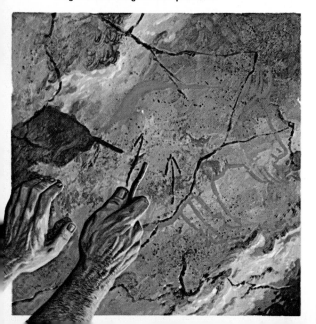

Top: In the Dark Ages organized science in Europe came to a halt. But in the Near East the alchemists – men who sought to turn base metals to gold, and to find an elixir which would give everlasting life – laid the foundations of modern chemistry by their observations on what happened when different substances were combined under different conditions. Alchemy spread to Europe and did not die out until well after the Renaissance. Above: The invention of the wheel in prehistoric times provided Man with an invaluable tool for technology and science.

Isaac Newton (1642-1727) was the first systematically to set out the laws of the universe. His most important work was done before he was 30 years old.

Above: In the 18th century science was a fashionable study, and society was intrigued by such experiments as this early demonstration of electricity.

Below: The development of the microscope in the late 1600s opened men's eyes to undreamed forms of life and a new understanding of the nature of diseases.

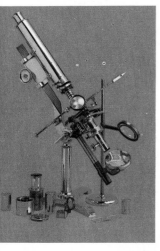

from the Church leaders; similarly they accepted the Greek scientific theories in an unquestioning way which would have appalled the Greeks themselves. But with the Renaissance came a new spirit of enquiry, and in the 1500s the great age of the challenge to authority. The power of the Church was defied by Martin Luther; and Nicolaus Copernicus flatly contradicted Ptolemy by saying that the Sun and not the Earth was the centre of the universe.

During the Renaissance learning became fashionable, and wealthy men prided themselves on their enquiring minds. The invention of printing brought the new ideas within the reach of scholars all over Europe, and the Church fought vainly to check this threat to its authority. At this time, too, new tools were invented which were an immense help to scientists. With his telescope Galileo discovered new heavenly bodies; with his microscopes Leeuwenhoek studied the teeming bacteria in a drop of water.

Leeuwenhoek's works were published by the Royal Society in London. Here, under the patronage of King Charles II, men met to discuss subjects ranging from 'valves in the veins' to 'the weight of air'. Its members included Robert Boyle, the eminent chemist, the astronomer Halley, and later Humphry Davy and Lord Rutherford, among many others. But most famous of all was Isaac Newton, whose laws of gravity and motion were the first really systematic setting out of the laws governing the universe.

The 18th century was the Age of Reason – and this was the ideal outlook for scientists. Enormous strides were made in chemistry and experiments with electricity began. The fashion for reason and order led to the classification of physical laws, chemical elements, and living things. Charles Darwin's theories of evolution in the 19th century were based on his observations and the classifications of earlier workers; in fact, he did not immediately publish his theories, and before he did so, another scientist reached some of the same conclusions.

The 18th century had been the age of the amateur scientist; and, as so little was known, it was still possible for one man to have a grasp of all the current knowledge in one quite broad field. But in the 19th century so many advances were made that science became the study of dedicated professionals in ever-narrowing fields. Today new tools – the radio telescope and the electron microscope – enable men to study farther horizons and minuter details than the scientists of even a hundred years ago would have dreamed possible. And from the investigation of one tiny part of a living cell – the gene – scientists may have made one of the most momentous of all discoveries – how life on Earth began.

But whatever the subject of study, the method of the scientist is unchanging: the observation and classification of facts, and finding their place in the order of nature.

A model of the double helix of DNA – the material from which genes are made. The discovery of this fascinating and complicated nucleic material may eventually lead to an understanding of how life began.

The Ancient World

Above: The cuneiform or wedge-shaped writing invented by the Sumerians. Writing enabled knowledge to be permanently recorded.

In the great agricultural civilizations of Mesopotamia and Egypt, science grew out of the practical problems of farming and building and was closely linked with religion.

Along the banks of the river Nile and in the land watered by the Tigris and Euphrates grew up two of the greatest early civilizations – those of Egypt and Sumeria, which flourished from about 3000 BC.

Science in Sumeria
We know from excavations that the Sumerians built such great cities as Ur and Babylon; to do so called for some knowledge of basic scientific techniques. They probably, for instance, used the wheel to transport the heavy building materials needed to construct the massive city walls. In order to lay out their cities and temples so accurately they must have understood the principles of simple geometry – for example, the properties of circles and rectangles. Their temples, for example, were usually built within an oval wall.

The Sumerians had other scientific skills than those needed for planning and building. They believed that the stars controlled the destiny of mankind, and this led them to observe the heavens and learn about the movements of stars and planets. Sumerian astrologers – men who predict the course of human affairs from the position of heavenly bodies – were therefore men of great power and importance. No man dared take a major decision, whether to wage war or to harvest the crops, without first consulting the astrologers on whether the time was right to do so. The astrologers were part priests, part scientists; science and superstition combined made up Sumerian religion.

The Sumerians were among the first to learn to calculate, and invented a symbol for the number 10 so that large numbers could be written more easily. It is to the Sumerians that we owe the division of the circle into 360 parts, or degrees, still used today.

The Sumerians' greatest contribution to science was writing. They invented a com-

Slaves carrying earth to make mud bricks for an Assyrian king's palace. Early civilizations had few mechanical aids; instead they depended on huge work forces.

plicated script which was pressed on to clay in cuneiform (wedge-shaped) symbols. Before written language, knowledge was passed down orally – by word of mouth – so much was lost. Knowledge could now be permanently recorded.

On the banks of the Nile
Life in Egypt was centred round the annual flooding and receding of the Nile. Each year boundaries were washed away and had to be relocated by taking measurements from permanent landmarks. To make these calculations and work out land areas the Egyptians invented geometry (from the Greek *geometria*, 'earth-measurement'). Their

The great civilization of Egypt developed from farming communities. The rich alluvial soil was worked by ox-drawn ploughs.

Below: The Great Pyramid – a monument to the skill of Egypt's architects. To construct the building, over two million 2½-ton stone blocks were somehow hauled across the sands and then positioned with unbelievable accuracy.

These simple tools, including a mallet and chisels, were used by the craftsmen and farmers of ancient Egypt.

geometry was so accurate that they built the Great Pyramid at Giza with a maximum error of only about a twentieth of a degree, and an astonishing uniformity in the length of its sides. They knew the properties of a right-angled triangle centuries before Pythagoras mathematically proved them. The Egyptians could also calculate the volumes of cones and truncated cones, and in this way they ascertained how much stone would be needed in the construction of their pyramid-shaped tombs. In astronomy the Egyptians made such accurate observations that they realized the year is $365\frac{1}{4}$ days long – although they continued to use a 365-day calendar. They believed that the Sun was the controlling influence on men's lives and, as in Sumeria, the magic of the priests and the science of the astronomers merged to become Egyptian religion.

The Egyptians were also advanced in the science of medicine. They disembowelled and mummified their dead, and in the process learned much about human anatomy. Their practical ability, however, was founded on superstition, and the belief that the god Horus looked after good health.

The Egyptians also recorded their knowledge, using pictorial characters, or hieroglyphs. It was this ability that set the Sumerians and Egyptians apart from others. Scientific knowledge could now be passed on to future generations.

An Egyptian fresco from a tomb at Thebes. It shows (top) the boundaries of a farmer's land being measured – an event that took place every year after the Nile's flooding had destroyed the old boundaries. The need for these annual land surveys led to the science of geometry. Below: The prehistoric monument of Stonehenge, England, may have been an astronomical observatory.

THE ROYAL CUBIT

Egyptian measurements were based on the *royal cubit*, the distance from the tip of the middle finger to the elbow (about 21 inches, 53 cm). This was divided into 7 *palms* (each about 3 inches, 7·5 cm) and 28 *fingers* (about $\frac{3}{4}$ inch, 2 cm). Area was measured in *setats* of 10,000 square cubits, and the unit of weight was the *deben* (about 3 ounces, 85 grammes).

Egyptian multiplication was very tedious, since the mathematicians were only able to double numbers and multiply by 10. Thus to multiply 3 by 18, for example, they would have to multiply 3 by 2 four times, and 3 by 10 once. The sum of all the products would provide the answer.

Natural Science and the Greeks

The philosophers of Ancient Greece speculated about the composition of matter and the structure of the universe, and tried to discover the laws of nature.

Above: The Greek philosopher Aristotle (384–322 BC) laid the foundations of many of the sciences of today by defining and classifying the various branches of knowledge.

Below: Part of the Acropolis in Athens. During Greece's Golden Age, many masterpieces of architecture were built. The Greeks were skilled builders who used their knowledge of basic geometry to create buildings of magnificent stature and elegance.

Most primitive peoples try to explain the workings of the universe in terms of the actions of gods. But the Greek thinkers wanted to explain things mechanically – leaving the gods out of it. Anaxagoras of Clazomenae boldly claimed that the Sun and Moon were nothing but lumps of rock, while Xenophanes of Colophon studied fossils and suggested that the Earth had once been covered with mud. Democritus and Heraclitus arrived by intuition at theories which were far in advance of their times. Democritus believed that matter was made up of tiny, indivisible components: he coined the term 'atom', meaning 'that which cannot be split'. Heraclitus asserted that everything was in constant motion – though the motion took place on such a small scale as to be invisible.

A centre of learning and culture

In the 5th century BC, the most illustrious period in Greek history, the centre of culture and learning was Athens. The Athenians loved to spend their spare time discussing science and philosophy, and many travelling teachers – called sophists – visited the city, giving lectures on such topics as mathematics, geometry, and astronomy.

In the 4th century BC the philosopher Plato made Athens the first university city in the world when he founded his Academy. His pupil, Aristotle, wrote a series of treatises on natural science, including two books on plants and another on the movement of animals. He put together what was then known of science and added some of his own ideas on biology.

In general the early Greeks were better at abstract thinking than at carrying out practical experiments and inventing labour-saving devices. This was partly because all Greek societies contained a slave class; the slaves did most of the hard work, leaving educated men free to ponder at leisure on abstract theories. But after mainland Greece had fallen under the dominion of Alexander the Great and his successors, and Alexandria in Egypt became the centre of Hellenistic (Greek-based) learning, there was a great explosion of practical knowledge. Euclid set out his theories of geometry which formed the basis of its study until the 20th century; Archimedes not only studied geometrical figures but set out mechanical principles and made many ingenious and practical machines, including a screw for raising water which is still used today. The Alexandrian astronomers included Aristarchus and Hipparchus, and their work culminated in that of Ptolemy. Medical knowledge reached a peak in the 2nd century AD, with the work of Galen.

The great Roman Empire provided peace and prosperity, under which learning could flourish. But as the Empire began to break up in the 3rd century, learning, too, declined. Europe was slipping into the Dark Ages.

Below: The frontispiece of the 1496 edition of Ptolemy's 'The Almagest', showing the author (left) and the German astronomer Johann Müller, known as Regiomontanus. Above them is an armillary sphere with the Earth in the centre and the signs of the Zodiac around it. 'The Almagest' outlined Ptolemy's theory of an Earth-centred universe.

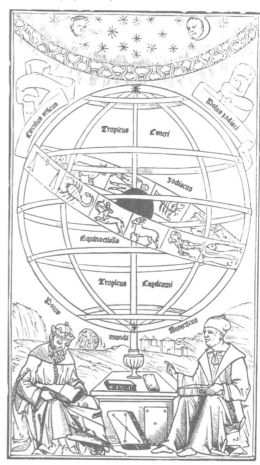

GREEK SCIENTISTS AND PHILOSOPHERS

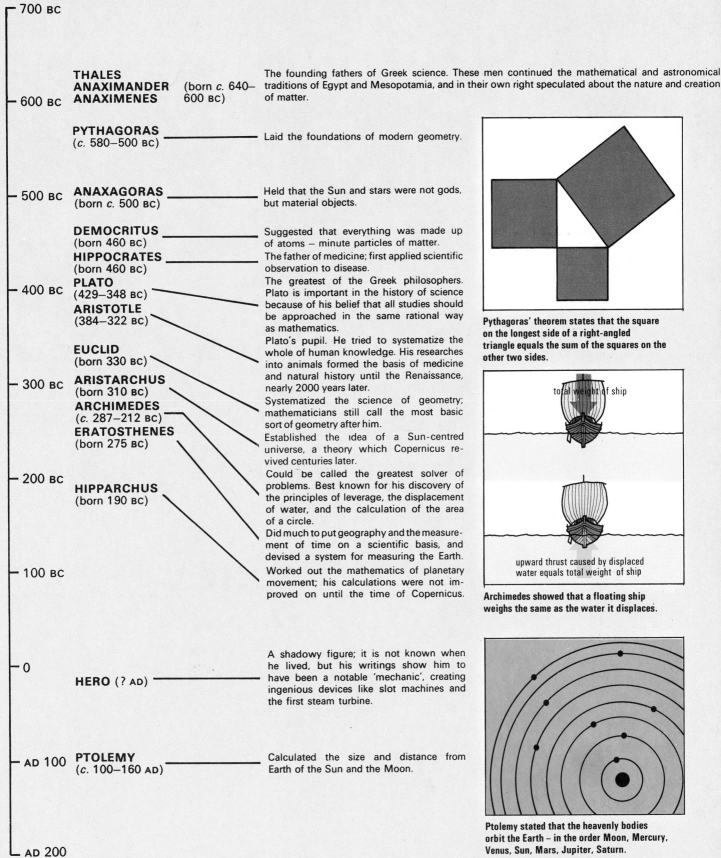

700 BC

600 BC

THALES
ANAXIMANDER (born c. 640–600 BC)
ANAXIMENES

The founding fathers of Greek science. These men continued the mathematical and astronomical traditions of Egypt and Mesopotamia, and in their own right speculated about the nature and creation of matter.

PYTHAGORAS
(c. 580–500 BC)

Laid the foundations of modern geometry.

500 BC

ANAXAGORAS
(born c. 500 BC)

Held that the Sun and stars were not gods, but material objects.

DEMOCRITUS
(born 460 BC)

Suggested that everything was made up of atoms – minute particles of matter.

HIPPOCRATES
(born 460 BC)

The father of medicine; first applied scientific observation to disease.

PLATO
(429–348 BC)

The greatest of the Greek philosophers. Plato is important in the history of science because of his belief that all studies should be approached in the same rational way as mathematics.

400 BC

ARISTOTLE
(384–322 BC)

Plato's pupil. He tried to systematize the whole of human knowledge. His researches into animals formed the basis of medicine and natural history until the Renaissance, nearly 2000 years later.

EUCLID
(born 330 BC)

Systematized the science of geometry; mathematicians still call the most basic sort of geometry after him.

300 BC

ARISTARCHUS
(born 310 BC)

Established the idea of a Sun-centred universe, a theory which Copernicus revived centuries later.

ARCHIMEDES
(c. 287–212 BC)

Could be called the greatest solver of problems. Best known for his discovery of the principles of leverage, the displacement of water, and the calculation of the area of a circle.

ERATOSTHENES
(born 275 BC)

Did much to put geography and the measurement of time on a scientific basis, and devised a system for measuring the Earth.

200 BC

HIPPARCHUS
(born 190 BC)

Worked out the mathematics of planetary movement; his calculations were not improved on until the time of Copernicus.

100 BC

0

HERO (? AD)

A shadowy figure; it is not known when he lived, but his writings show him to have been a notable 'mechanic', creating ingenious devices like slot machines and the first steam turbine.

AD 100

PTOLEMY
(c. 100–160 AD)

Calculated the size and distance from Earth of the Sun and the Moon.

AD 200

Pythagoras' theorem states that the square on the longest side of a right-angled triangle equals the sum of the squares on the other two sides.

total weight of ship

upward thrust caused by displaced water equals total weight of ship

Archimedes showed that a floating ship weighs the same as the water it displaces.

Ptolemy stated that the heavenly bodies orbit the Earth – in the order Moon, Mercury, Venus, Sun, Mars, Jupiter, Saturn.

Light in the Dark Ages

*Science and learning were almost lost to the West
when the Roman Empire fell to the barbarian hordes.
But the progress of knowledge continued in the
East — in India, China, and, later, the Arab world.*

In the 4th century BC the city of Alexandria in Egypt became the centre
of Greek learning, and over the next 300 years its Museum grew into a
treasure house of ancient knowledge. At its peak the Museum's great
library contained some 70,000 volumes. When the city was besieged by
the Romans in 47 BC the library was destroyed, but a 'daughter library'
grew up, based on manuscripts given by the Roman Mark Antony.
This was also destroyed, on the orders of Emperor Theodosius I, in AD 390,
and with it was lost centuries of hard-won knowledge.

At that time the Roman Empire was suffering barbarian invasions.
In 410 the Goths captured Rome, and the city was plundered by the
Vandals in 455. The great Roman civilization had slowly crumbled
away. With the disappearance of law and order, the trade routes between
East and West became dangerous and unprofitable. The scholars who
had travelled with the traders' caravans now stayed at home, and the
interchange of ideas and scientific knowledge between the two worlds
dwindled. Western science had suffered a crushing set-back.

To the east, however, remained the great civilizations of China and
India. Here there was no such break in learning.

The science of the East

The civilization of China, founded on the banks of the Yellow River, was
now about 2000 years old. Early Chinese civilization was based on agri-
culture and so the calendar was important; as far back as the reign of
Wu-ting (1324–1266 BC) they had a satisfactory, if cumbersome, year of
12 months of 30 days each, with an extra 13th month added when
necessary. They took an intense interest in the heavens and had mapped

Above: In 410 AD Rome was sacked by Alaric and his
Goths. The great Roman Empire had provided stable
conditions under which science flourished, but when it was
overrun by the barbarians its cities were burned, its libraries
destroyed, and the struggle for survival left no time for
abstract thought. Below left: Long before the telescope the
ancient astronomers were able to make astonishingly
accurate observations of when particular stars were in a
particular spot in the heavens. During the Dark Ages
eastern scientists used great transits, built of masonry, to
observe the stars and determine the key points in their
calendars. Below: A Chinese water clock dating from the
11th century. The Chinese civilization suffered no
catastrophic break like the European Dark Ages.

nearly 1500 stars by the 4th century BC; they listed eclipses, comets, and meteors. Between AD 249 and 420 they noted 28 solar haloes (circles of light around the Sun) – phenomena not recorded in Europe until the 17th century. They were also skilled mathematicians and doctors. But Chinese science, like the rest of Chinese culture, remained almost completely cut off from the West.

Indian science, too, placed great emphasis on astronomy and medicine. But unlike that of China it made an immensely significant contribution to the Western world. For it was in India that the decimal system and the use of nine figures and a zero were developed.

Islamic contributions

In the 7th century AD the successors of the Arab Muhammad, the founder of Islam, built up an empire which was to stretch from Spain to India. Muhammad's followers, the Muslims, eagerly pursued his instruction 'to seek knowledge from the cradle to the grave'. About the middle of the 9th century, translators in Baghdad began to produce Arabic versions of translations by Syrian Christians of Greek scientific works. At about the same time the Arabs adopted the Indian number system and the use of what are commonly called Arabic numerals. These made calculations immensely easier than the old systems such as the Roman, with its use of letters for numerals and lack of a place system. The Arabs were very skilled mathematicians.

The Muslims turned five times a day to pray in the direction of Mecca. To do this they had to work out from the stars the direction in which Mecca lay, and several great observatories were built for this purpose. They were keen astronomers, and their major book of reference was the *Almagest*. This was in fact the Arab name for the astronomical work by Ptolemy of Alexandria.

In the same way, the Arabs used Greek works as a basis for their knowledge of medicine. Their greatest physician was the Persian Rhazes (*c.* 865–923), whose great work *Al-Hawi* surveys Greek, Syrian, and even some Hindu medical knowledge as well as opinions based on his own wide experience.

As the Western countries became civilized once more, they became increasingly interested in Arab learning. Works on astronomy and many of the treatises of Rhazes were among the books translated from Arabic to Latin. So knowledge based on the long-forgotten work of the Greeks, in many cases improved and expanded by the Arabs, took hold once more in Europe.

Above: According to tradition the Chinese knew in the 3rd millenium BC that a lodestone automatically points to the magnetic north or south, and by about 800 AD were using it for navigation. In the late 12th century Chinese floating compass-needles were introduced into Europe. This Italian miner's compass dates from the 1600s. Below: Arab astronomers. The Arabs translated works by the ancient Greeks and added their own observations.

The Alchemists

The alchemists' attempts to turn base metals into gold were doomed to failure. But their experiments were not entirely wasted – their observations and methods laid the groundwork for modern chemistry.

Earth, air, fire, and water: these, Aristotle stated, were the four elements from which all substances were made up. The differences between substances were caused simply by the different proportions of elements. So it was not very extraordinary for people to think that they could alter one substance into another: all they had to do, they reasoned, was find out just how they could alter the proportions of elements. And if they could change soft, grey lead into gold, most precious of metals, then the result would be immense wealth and power.

Flasks and furnaces

The people who sought to alter substances were known as alchemists. They probably started work in the Egyptian city of Alexandria at about the time when Christ was born. Alexandria was a great trading city and people from all nations mingled there, so it was not surprising that alchemy combined Greek ideas on matter with Mesopotamian belief in astrology. Their

Above: This painting by Joseph Wright shows the German alchemist Hennig Brand in his laboratory, bathed in the light of the element phosphorus which he discovered in 1669. Alchemy formed the basis for modern chemistry.

'science' was an extraordinary mixture of practical experiments and mystical ideas; the flasks and furnaces they designed were decorated with secret symbols and they invented elaborate relationships between the metals and the stars.

One of the alchemists' great goals was to find the 'philosopher's stone', a substance which would change any metal into gold simply by touching it. They also sought the *magisterium* which would cure all diseases. If they could only discover this, they thought, it would be but a small step to the *elixir vitae* – the elixir of life – by which a man could prolong his life for as long as he wished. In their experiments they melted and combined and distilled all manner of substances and gained considerable knowledge of chemical reactions.

From alchemy to science

Alchemy flourished in Europe and the Arab world. Many alchemists were quacks and charlatans, always ready to cheat the

---

JABIR AND GEBER

Some 2000 early books on mystical subjects including alchemy are attributed to an Arab alchemist, Jabir ibn Hayyan, who is thought to have lived in the 8th century AD. Whether or not Jabir existed and whether he wrote all the books is not known, but the works strongly influenced alchemists who read them. The mysterious Jabir gained such a reputation that by the 14th century a Spanish alchemist had adopted the westernized form of the name, Geber, to give his work more authority. Geber's own four books gave a clear description of the aims and methods of alchemy, and did much to improve its standing as a science. For over 200 years they remained the most authoritative books on alchemy.

A design for a 'perpetual motion' device – one that would move for ever without energy from outside. No one has yet invented such a machine; the British Patent Office refuses applications on the grounds that such a device is contrary to well-established physical laws.

THE QUEST FOR PERPETUAL MOTION

At the same time that alchemists in the Middle Ages were seeking to turn base metals into gold, other men were trying to perfect a machine which, once started, would continue working for ever without using any power.

There have been many attempts to construct a perpetual motion machine; one of the first such devices was built in the 13th century by the architect Villard de Honnecourt. None has been successful. From the quest for this machine, the law of conservation of energy was born. This states that energy cannot be created or destroyed, it can only be changed in form. The law rules out the possibility of perpetual motion because the energy which is unavoidably lost through friction must be replaced by an outside source of power.

In the late 19th century the American John Keely claimed to have constructed a perpetual motion motor. He gave fraudulent demonstrations, talking of his 'hydro-pneumatic-pulsating-vacuo-engine', and made large sums of money before his trickery was exposed.

Roger Bacon was accused of black magic, and confined for ten years by his Franciscan order because of his 'heretical' writings.

An alchemical tree, showing some of the strange symbols. The alchemists were obsessively secret in their search for the philosopher's stone, writing in code and jealously guarding their findings.

gullible public. But others were serious men of science. Among these was Roger Bacon (?1220–1294). Bacon expended time and energy and huge sums of money on research; he was interested not only in alchemy and astronomy but also in mathematics and optics. He became a Franciscan friar and with the encouragement of the Pope wrote a great encyclopedia covering the sciences of his time.

Although Bacon was a practising alchemist, he went about his experiments in a very modern way. He believed that truth is found by observing how things behave rather than accepting old theories. He also described what may have been a kind of microscope and showed how lenses could be combined to make objects appear nearer. He constructed a kind of primitive steam engine and even foretold of a time when ships and carriages would be driven by engines, and men would fly through the air in machines.

The search for the philosopher's stone and the elixir of life was in vain, and the new ideas and learning of the Renaissance swept away the mysticism of the alchemists. But the knowledge of chemistry that they had gained, reinterpreted by the new rational thinkers, was of incalculable value.

The World Reborn

Above: A self-portrait by Leonardo da Vinci (1452–1519). Possibly the greatest man of the Renaissance, he was both a brilliant painter and a scientific genius.

As Europe emerged from the Middle Ages, a new spirit of enquiry inspired the arts and sciences. A genius of the period was Leonardo da Vinci, who did brilliant work in both fields.

The word *renaissance* means re-birth; as used today it generally refers to the revival of classical learning that started in Italy in about the 14th century and spread all over Europe. Artists began to work directly from nature and to study mathematical proportion, as had their Greek and Roman forebears. Writers tried to imitate the high style of classical authors in modern languages. With the Renaissance in the arts came the Renaissance of science. But this was not a revival of classical science; in some cases it was even a rejection of classical learning as people began to believe the evidence of their eyes rather than existing authorities, however distinguished.

An age of enterprise
At the same time, the way people lived was changing. The old feudal system was dying out and a new class of traders and artisans, living in towns and respecting intelligence and enterprise rather than birth, was growing. These people were behind the rapid development of technology in this period, as they searched for more efficient ways to manufacture goods. The rise of great mining families in northern Europe meant advances in assaying – the art of analysing ores and mixtures of metals – which laid some of the foundations of the science of chemistry. Another invention which had an important indirect effect on science was printing with movable type; this was essential to the spread of ideas from one country to another. These technological advances came mainly from northern Europe, unlike the Renaissance in the arts which flowered in Italy. But one of the greatest men of the scientific Renaissance, who embodied both its questioning spirit and its inventiveness, was an Italian artist, Leonardo da Vinci.

Leonardo, born in 1452, was not simply an artist who was interested in science; both ways of thinking were united in his character. His notebooks contain drawings of countless projects and their explanations, set down in mirror writing for secrecy. They range over such subjects as anatomy, civil and military engineering, flight, and the movement of water. Again and again the links between the artist and the scientist can be seen. Leonardo's paintings show his interest in effects of light and colour; his notebooks record his scientific studies of light. He seems to have been the first person to notice that shadows are not black, but blue-grey. The study of light led him on to study the eye itself. Another scientific study that arose from the artist's practical needs was anatomy; in this Leonardo was a pioneer. He was one of the first people to dissect human corpses to learn about the structure of the body under the skin; this practice brought him trouble with the

Below: The Flemish city of Bruges in the 15th century, where increased trade led to the introduction of machines – with their extra efficiency and speed.

Below: Saint Peter's basilica in Rome is an example of the Renaissance throwback to classical architecture. The dome, for example, is built to precise mathematical proportions – typical of Greek and Roman building.

Church. He left hundreds of accurate drawings of such things as the construction of the brain and the growth of an unborn child in the womb.

Leonardo the prophet

Leonardo despised anatomists who could not draw and so record the progress of their dissections. His own talent for drawing was extraordinary – he could capture the most minute and fleeting details on paper. He even made a series of drawings of water in motion, which was the only record of the subject until the invention of high-speed photography. Only a brilliant artist could draw so accurately; only a scientific mind would bother to try.

Leonardo also employed his talents in the design of machines. He is credited with many lasting inventions, such as pointed bullets, the paddle wheel for boats, and the mincing machine. But it is as a prophet of modern technology that Leonardo the scientist is best remembered today. He anticipated such modern ideas as the tank and the parachute; and he was fascinated by the idea of a flying machine. As an anatomist, Leonardo knew human muscles alone could not power flight as those of birds do, so he explored ways of flying helped by currents of air. But the wings of his aircraft are all based on careful study of the wings of birds and bats. None of these machines would have flown if ever built – but nobody since his time has succeeded in achieving sustained, powered flight without the enormous power of the internal combustion engine.

Leonardo left many of his paintings and other projects unfinished – frustrated, it seems, by the impossibility of reaching absolute perfection. The world in the Renaissance was already too complex for one mind; scientists and other thinkers were becoming specialists, restricting their studies to ever-narrower fields.

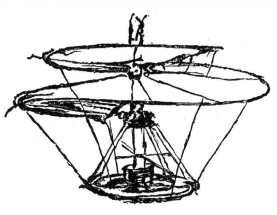

Above: Leonardo was the first man to design a helicopter. In this he was centuries ahead of his time – the first manned helicopter did not fly until the early 20th century.

Right: A striking example of the fusion of Leonardo's artistic and scientific abilities. His anatomical drawings were extremely accurate and easily surpassed the work of his contemporaries.

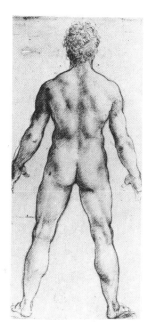

Below: One of Leonardo's many flight machines.

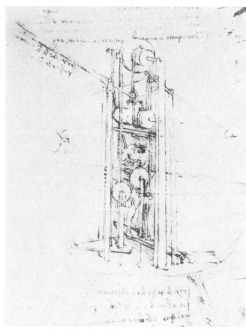

A New Look at the Universe

Copernicus' statement that the Earth moved round the Sun was vehemently denied by the Church authorities. But he was right — and his 'heresy' proved the key to the understanding of the universe.

Above: Nicolaus Copernicus (1473–1543), who established that the Sun, and not the Earth, was the centre of the universe.

Right: Galileo with the telescopes he used to observe the Sun and planets.

In 1492 Columbus landed in the Americas. By sailing west he had reached what he thought was the East; and the idea that the Earth was round gained in popularity. Among the people whose imagination was fired by Columbus's voyage was a 19-year-old student from Cracow in Poland, Nicolaus Copernicus (1473–1543).

Copernicus studied the works of the ancient Greek scholars, who had also believed the Earth was round. When he considered the work carried out by these men, he came across another theory put forward by Aristarchus nearly 18 centuries earlier but long discarded: that the Sun was the centre of the universe.

Like many Renaissance scientists, Copernicus was wealthy enough to pursue a life of study. Educated first at Cracow, he moved on to the great Italian universities at Padua, Bologna, and Ferrara.

Below: The Solar System as seen by Copernicus.

Above: Johannes Kepler (1571–1630) proved that planets orbit in ellipses rather than circles.

Below: Isaac Newton (1642–1727), one of the most brilliant of all scientists. His theory of gravity is still accepted after nearly 300 years.

By 1536 Copernicus had almost finished his great work, *Concerning the Revolutions of the Heavenly Spheres;* but it was not published until the year of his death, aged 70, in 1543. In it Copernicus maintained that the Sun was the centre of the universe and that the Earth was only one of the planets that revolve around the Sun.

Planetary motion

Tycho Brahe, a Danish astronomer born three years after the death of Copernicus, was a believer in Ptolemy's theory of an Earth-centred universe. His great contribution lay in detailed and accurate observations, rather than new theories. When he died in 1601, he left the whole mass of his very accurate and detailed work to his successor Johannes Kepler, and with the aid of Brahe's observations Kepler was able to investigate mathematically the theories of Nicolaus Copernicus. It was Kepler who found that planetary orbits are ellipses, rather than circles, as Copernicus had thought, and it was Kepler who expressed the laws of planetary motion, which paved the way for the work of Sir Isaac Newton.

Roughly contemporary with Kepler was the brilliant scientist, Galileo Galilei. As a professor of mathematics at the University of Pisa he challenged Aristotle's teaching that heavy bodies fall faster than light ones. Galileo reasoned that the force of gravity pulls all bodies to Earth with the same acceleration, regardless of their weight.

In 1609 Galileo built his own telescope

and began observing the planets and stars. He became convinced that Copernicus' Sun-centred theory of the universe was correct. He demonstrated some of his discoveries to the Pope in Rome, but the Church bitterly opposed his theory that it was the Earth that moved round the Sun and he was ordered to recant. The story goes that having done this, Galileo added under his breath, *'E pur si muove'* ('Yet it *does* move'). Galileo's life's work on gravity, acceleration, and the laws of motion was set out in his *Dialogues on the Two New Sciences*, published in 1638.

A universal force

Sir Isaac Newton was born in 1642. He was a mathematician, astronomer, physicist, and, as he would have been termed in his own time, a natural philosopher — a man who thought deeply about all that he observed in the universe around him.

As a young man, Newton mastered the science of mathematics, and added to it his 'invention' of calculus, a branch of mathematics. Before his 30th birthday he had conducted many experiments to determine the nature of light, and invented the type of reflecting telescope that bears his name.

Newton recognized that there was a force that extended throughout the universe, and controlled the stars in their courses as much as it did the fall of an apple from a tree — the universal force of gravitation. He showed that the gravitational force of the Sun keeps the planets in their orbits. Newton's laws of gravitation and motion were published in 1687 in *Philosophiae Naturalis Principia Mathematica* (Mathematical Principles of Natural Philosophy). It was the first book to set out a unified system of scientific principles explaining how the universe is held together, and it became the bible of physicists for nearly two centuries.

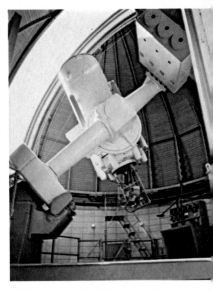

Above: A modern 41-inch reflecting telescope, at Yerkes Observatory, USA.

Below: One story has it that Galileo proved that all objects fall with the same acceleration by dropping two different weights from the top of Pisa's Leaning Tower.

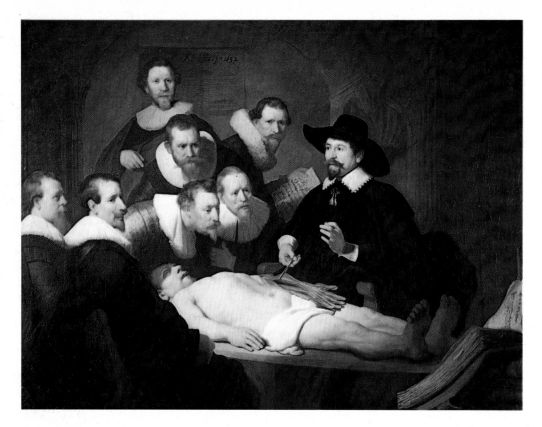

Rembrandt painted 'The Anatomy Lesson' in 1632. For a long time the Church had strongly opposed the dissection of corpses, since it regarded the human body as inviolable, and the early anatomists had often been forced to carry out dissection in secret. But the tide of the Renaissance was too strong, and by the mid 17th century anatomy was an accepted study.

Below: A portrait of Vesalius with his anatomical models. Vesalius first learned anatomy by dissecting the bodies of dead criminals. As well as attempting to standardize anatomical terms – and his nomenclature forms the basis for the terms used today – his works set a new standard of clear and even beautiful medical illustration.

The Anatomists

The Church authorities had rejected Copernicus' vision of a Sun-centred universe. Vesalius and Harvey were to face similar religious opposition as they fought to disprove many of the prevailing ideas about the human body.

In the 2nd century AD, Claudius Galenus, better known as Galen, a physician from Pergamum (now Bergama, in Turkey), discovered that arteries contained blood and not air as some Greek physicians had taught. He also knew that in some way the heart made the blood move, but he did not understand how it circulated. Galen believed that blood behaved like the tides, reaching all parts of the body with a gentle ebb and flow.

The founder of modern anatomy

This theory, and the belief that different parts of the human body were controlled by different stars, were widely accepted until the time of Andreas Vesalius, an anatomist born in Brussels in 1514. The new knowledge that grew out of the Renaissance destroyed many old ideas, among them those of the close links between Man – the microcosm – and the Universe – the macrocosm. Vesalius was the founder of modern anatomy. In 1538 he published *Tabulae anatomicae sex*, an attempt to standardize anatomical terms, on which we base our present terminology. His most important work came later in 1543, when, following numerous dissections, he published his masterpiece, *De Humani Corporis Fabrica*, the first complete description of human anatomy. Galen's supporters angrily reproached Vesalius for contradicting Galen's theory of the movement of blood, although Vesalius had not proposed his own theory. He became so discouraged by the poor reception of his study that he burnt most of his work.

In 1564 Vesalius was condemned to death by the Church authorities on uncertain grounds — either for stealing bodies from graves and dissecting them, or for carrying out a post mortem on a man whose heart

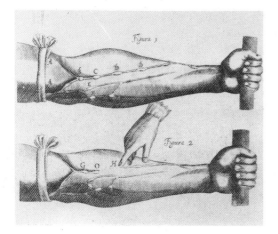

Above: A drawing from Harvey's great work on the circulation of blood, published in 1628. Scientists sometimes employed artists to draw illustrations for them.

was still beating. The sentence was commuted to a pilgrimage of penance to Jerusalem, but Vesalius died on the return journey at the age of 50.

Understanding the heart
Vesalius had made an immense contribution to the field of anatomy, but mystery still surrounded the action of the heart. The English physician, William Harvey, provided the solution.

Harvey was born at Folkestone, England, in 1578. At the age of 16 he went to Caius College in Cambridge to study medicine. John Caius, a founder of the college, had himself been a pupil of Vesalius at Padua in Italy, and the young Harvey went there to continue the studies he had started at Cambridge.

Harvey lived in a world in which machines were beginning to do Man's work for him. He attended lectures in Padua given by Galileo at a time when the great Italian was investigating the mechanics of the pump. It may be that Harvey was influenced by Galileo's work, for he later likened the action of the heart to that of a pump.

On his return to England, Harvey both practised and taught medicine. His practice

flourished, and he became physician to two kings, James I and Charles I. At the same time he continued his work on the movements of the heart and blood. Harvey measured the amount of blood which the heart pumped at a single beat — about one sixteenth of an ounce — and calculated that the heart's output in about half an hour must equal the total amount of blood in the body. He found that the blood moved away from the heart in arteries and towards it in veins. The remaining question was how the blood passed from the arteries to the veins. Harvey guessed at the presence of capillaries — tiny blood vessels which connect the smallest veins and arteries. All his theories were gathered together in one of the most important works in the history of physiology, *An Anatomical Treatise on the Movement of the Heart and Blood in Animals*, published in 1628. Harvey lived to see his theories accepted by eminent anatomists all over Europe, where he was known and honoured.

Despite strong opposition from the Church, which regarded dissections as a violation of the human body, Vesalius and Harvey had radically changed people's ideas about the human body, and provided the basis for modern anatomy.

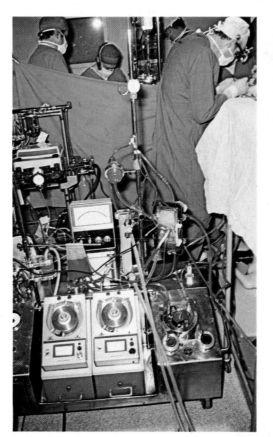

Above: A drawing by Vesalius showing the major muscles of the human body. Vesalius was the first man to write a complete description of human anatomy.

Above: William Harvey (1578–1657) proved that blood circulated around the body and conceived the notion of the heart as a pump.

Left: Our knowledge of human anatomy is such that we can now build machines that serve as parts of the body. This heart/lung machine functions as the heart and lungs during an operation. Without such machines it would not be possible to perform certain types of open heart surgery.

Invisible Elements

The pioneer chemists of the 17th and 18th centuries were fascinated by the most familiar but intangible of substances — air.

In the 17th century, air — one of the mystic elements of Aristotelean science — was losing its mystery. The invention of the air pump by Otto von Guericke in 1650 enabled Robert Boyle, an Irish chemist, to investigate the properties of gases. In 1662 he published what is now known as Boyle's Law, which says that the volume of a gas at a constant temperature varies inversely with the pressure upon it.

In the same year Boyle made another important discovery. While working in his laboratory he noticed that a candle burning in an enclosed space is extinguished before all the air has been used up. He concluded that air consists of two parts, one which supports combustion and one which does not. This theory contradicted the belief held since the time of Aristotle that air, fire, earth, and water were the fundamental elements.

Above: Boyle's Law. If the pressure on a gas is doubled, its compressed molecules (represented here by lines) occupy half the space.

JOHN DALTON 1766–1844 first introduced the idea that all compounds consist of molecules containing elements in a fixed proportion.

Left: Charts of the elements and their atomic weights and models of atoms, used by John Dalton in his lectures.

Phlogiston

Following Boyle's work on burning substances, the German chemist, Georg Ernst Stahl, put forward a theory of the burning process in 1697. He claimed that all inflammable materials contained 'phlogiston' and that this was given off when they burned, and carried away by the air. His theory seemed to reinforce Boyle's beliefs about air, since combustion would stop when all the phlogiston in a substance was given off, though air might still be present.

Most scientists seemed to regard all gases as some form of air. In 1756, Joseph Black, a Scottish chemist and physician, discovered carbon dioxide and called it 'fixed air'. Henry Cavendish discovered 'inflammable air' in 1766. We now know it as hydrogen.

Joseph Priestley (1733–1804), an English chemist, did not share this belief that all gases were a form of air. He discovered carbon monoxide, hydrochloric acid gas, and sulphur dioxide. In 1772 Priestley and Daniel Rutherford independently discovered nitrogen, the gas which makes up 78 per cent of the air. But in 1774 Priestley made by far his most important discovery. By heating mercuric oxide, he obtained a new gas which appeared to support combustion far better than ordinary air. He noticed that mice became very frisky when breathing this gas, and that he himself felt 'light and easy' in its presence. Priestley had discovered oxygen. The part that oxygen plays in combustion was first understood by Antoine Lavoisier, a French chemist.

The secret of burning

Antoine Laurent Lavoisier (1743–1794) was, with Robert Boyle, the founder of modern chemistry. He correctly explained combustion as a combination of the burning substance with Priestley's new gas, which he now termed oxygen. Lavoisier showed that when a metal is heated in an enclosed space, the increase in weight of the metal is exactly equal to the loss of weight of the air in the enclosure. Lavoisier's experiments led him to formulate, in 1789, the law of conservation of matter, which states that matter can be neither created nor destroyed.

Lavoisier has often been criticized for not giving credit to scientists like Priestley and Cavendish whose work he built upon, but he remains one of the greatest brains of the 18th century. When he was guillotined in 1794 during the French Revolution on a petty accusation, a former colleague remarked, 'It required only a moment to sever that head, and perhaps a century will not be sufficient to produce another one like it'.

Air and other substances had now been broken down into their constituent elements. It remained for John Dalton, an English chemist, to show how the parts were atomically bonded together.

John Dalton (1766–1844) took up teaching when he was only 12 and studied science in his spare time. He discovered that when two gases combine they always do so in simple, fixed proportions. Dalton's famous atomic theory, which appeared between 1808 and 1827, stated that an element consists of identical atoms which always weigh the same, and that these combine in simple proportions to form chemical compounds. His principles have been generally accepted as the basis of modern atomic theory.

ROBERT BOYLE 1627–1691 (top left) was a founder-member of the Royal Society, whose object was 'improving natural knowledge by experiment'.

JOSEPH PRIESTLEY 1733–1804 (right) is best remembered for his discovery of several gases, notably oxygen.

ANTOINE LAURENT LAVOISIER 1743–1794 (right) published what was perhaps the first modern chemistry book, 'Elements of Chemistry'.

Below: Priestley works on the problem of combustion.

The Ladder of Life

When it became clear that the Earth was not the centre of the universe, scientists began to question the unique place of Man in the scheme of creation.

Above: The biblical story of creation was widely believed to be literally true until the middle of the 19th century.

Men of all religions and creeds had always believed in the theory of 'individual creation' – that all living things had been created separately and individually by some divine being, whatever name was given to him. Man was the creator's greatest achievement.

When work began on the science of comparative anatomy – comparing the ways in which different creatures were constructed – this belief was shaken. In the 18th century men like the Scottish anatomist John Hunter and the pioneer embryologist Karl von Baer pointed out such disturbing facts as the resemblance of the embryos of higher animals like Man to the adult forms of lower ones such as fish. They also showed that the things that mark the difference between species develop later in embryos than the things they have in common. For example, the embryo of a mammal begins to look like a mammal only towards the end of its growth. Could the theory of individual creation be wrong?

'The Ladder of Nature'

When the Swedish naturalist, Carolus Linnaeus (1707–1778) introduced his system of classifying plants and animals, he called Man 'a master animal'. In his system, still used today, each living thing has a two-part Latin name: the first is the *genus* (group) and the second is the *species* (kind). For example, the dog-rose is *Rosa canina*. Linnaeus believed that his classification of the natural world applied to all periods of time. While claiming that 'Man is the only creature that God has created with an immortal soul', he had to admit that he found it difficult 'to discover one attribute by which Man can be distinguished from apes. . . .'

As a result of Linnaeus's work the distinctions between living things became easier to pinpoint. The French naturalist, Georges Cuvier (1769–1832), who also believed in individual creation, classified the animal world with such completeness that he could virtually identify an animal from a mere fragment. But there was still great ignorance surrounding the animal kingdom even in the august world of the French Academy, that guardian of the purity of the French language. In 1799, when discussing the definition of a crab, someone proposed, 'A small red fish that walks backwards'. Asked for his opinion, Cuvier replied: 'Your definition is excellent, except for the fact that a crab is not a fish, it is not red, and it does not walk backwards.'

At about the time Cuvier was producing his systematic work, the Chevalier de Lamarck (1744–1829), was putting forward sensational theories on the development of species. He visualized a 'ladder of nature', in which each species had been changed into another, and the extinct ones had given rise to those now living. In a more limited way, this had already been

SEVERINO 1580–1656 One of the pioneers of comparative anatomy. In 1645 he began to record similarities among more and more widely differing creatures.

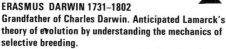

GEORGES CUVIER 1769–1832 French naturalist. Produced the first complete classification of the animal kingdom.

JOHN HUNTER 1728–1793 Appointed surgeon extraordinary to George III in 1776. Noticed that the embryos of mammals pass through stages of resemblance to lower animals.

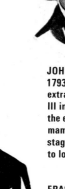

ERASMUS DARWIN 1731–1802 Grandfather of Charles Darwin. Anticipated Lamarck's theory of evolution by understanding the mechanics of selective breeding.

CHARLES DARWIN 1809–1882 Great English naturalist. Proposed the theory of natural selection, the first coherent theory of evolution.

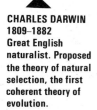

GREGOR MENDEL 1822–1884 Austrian botanist who formulated the basic laws of heredity through his breeding experiments with peas.

appreciated by the British scientist Erasmus Darwin (1731–1802), grandfather of the great Charles Darwin. He realized that the age-old farming practice of selective breeding improved the quality of farm stock through controlling the inheritance of the best characteristics of the species.

Voyage on the 'Beagle'

Charles Darwin (1809–1882), was the first person to put forward a sound theory of

evolution backed by observation. As a young naturalist on a five-year-long survey of the South Atlantic and Pacific Oceans aboard the *Beagle*, Darwin's ideas took shape. He noticed that the giant tortoises of the Galapagos Islands differed slightly from one island to another. Environment, he concluded, had put its mark on each separate genetic line. The same was true of the birds he found there. In 1859 he startled the world with the publication of his famous book, *The Origin of Species*, in which he set out his theories of natural selection, or the 'survival of the fittest'.

Darwin stated that if an individual happens to have different characteristics from its relatives, and if these characteristics enable it to survive more easily, it will thrive, while the others will die out. In this way a superior species will evolve.

The secret of heredity

An Austrian monk named Gregor Mendel (1822–1884) formulated the basic laws of heredity, or how certain characteristics are passed from parents to children. This was in 1864, following a series of experiments on the cross-breeding of different species of peas. What he had discovered was the existence of genes, which carry information from one generation to the next. Mendel's theory that changes in the form of an animal were due to cross-breeding between different forms of the animal seemed at the time to contradict Darwin's natural selection theory. It was to be nearly a century before new discoveries in the field of genetics brought the theories of Darwin and Mendel together.

In 1831, Charles Darwin was appointed naturalist to HMS 'Beagle', a survey ship due to make a voyage around the world. In the Galapagos Islands he noticed how the tortoises (left) and finches (right) differed from island to island. He realized that this was due to the different environments in which they lived. On his return to England he started work on 'The Origin of Species' which set out his theory of evolution.

Trofim Lysenko, born in Russia in 1898. In 1928 he started to make a series of claims that contradicted Mendel's views on genetics. For example, he stated that a plant's characteristics could be permanently changed by grafting. He was strongly supported by Joseph Stalin and made a member of the Supreme Soviet. His views were enforced on Soviet scientists. After Stalin's death in 1953 he fell from favour and was publicly discredited in 1965.

In 1857 Gregor Mendel began a series of experiments with garden peas. He cross-bred green (g) peas with yellow (Y) peas and saw that the next generation peas were all yellow. But when these peas were cross-bred in turn, they always produced one green pea for every three yellow. Mendel concluded that the first generation of 'hybrid' peas carried the traits of both parents, but that the Y or dominant trait over-ruled the g or recessive trait. From his experiments on peas Mendel formulated his Laws of Heredity.

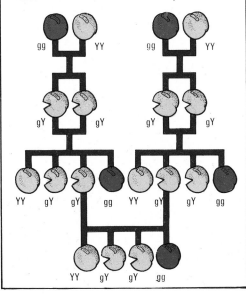

gg   YY   gg   YY

gY   gY   gY   gY

YY gY gY gg   YY gY gY gg

YY gY gY gg

Tools of the Scientist

Above: An Egyptian balance, a reasonably accurate measure for grain and other dry goods. The Egyptian unit of weight was the deben, which was about 3 ounces (85 grammes).

Through the ages, scientists have demanded precise tools and instruments to examine, measure, and record natural phenomena. Inventions like the telescope and microscope have revealed vast new areas for scientific research.

The scientist's need for tools to detect and measure physical phenomena is reflected in the early appearance of the telescope, the microscope, and other devices that had an immediate impact on developments in science. Man has always used tools to help him in his work. A scientist's tools differ from others in their precision. The Egyptian pyramids and Stonehenge, for example, could only have been constructed so accurately with the aid of scientific instruments to measure lengths and angles.

The 17th century was the great period of invention. One of the first telescopes was made by the Dutchman Hans Lippershey in 1608, when comparing two spectacle lenses, one behind the other. The telescope enabled men to view the heavenly bodies with much more accuracy, and led directly, for example, to Galileo's discovery of the four bright satellites of Jupiter.

A whole new world of study was opened to scientists by the invention of the microscope by another Dutchman, Zacharias Janssen, in 1590, although it was not widely used until Robert Hooke developed it in the 1660s. Now hitherto invisible objects — the hairs on an insect's leg, the cells in a leaf, the micro-organisms in the air — could be studied.

The chronometer, a very accurate time-keeping device, was first constructed and successfully used by an Englishman, John Harrison, in the 18th century. This instrument enabled men to plot accurately their course at sea, from taking timed observations of the Sun and stars.

The huge and complex atomic particle accelerator which scientists use today may be a far cry from the simple Egyptian weighing balance, but both were invented as tools to help the scientist to observe and record facts and build up a picture of the universe.

Above: A 16th-century print depicting the Greek scientist Ptolemy, guided by the Muse of Astronomy, using a quadrant, an instrument long known to astronomers, to measure the altitude of the Moon.

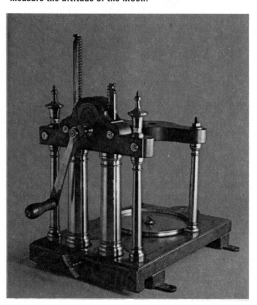

Left: The vacuum pump, first developed in the 17th century, was used by Otto von Guericke and Robert Boyle in their early experiments with air.

Right: An early electrical machine. These simple devices were used for producing electricity in the laboratory in order to carry out experiments in electromagnetism.

Above: An analytical balance patented by F. Sartorius in 1870. It can detect a difference in weight of one-tenth of a milligram, an outstanding performance for an instrument of its day.

Below: 'The Orrery', by Joseph Wright. These working models of the planets in the Solar System were very popular in the 18th and 19th centuries.

Above: A microscope belonging to Robert Hooke, whose great work 'Micrographia' (1665) described in detail the structure of the plant and animal cells he observed. Hooke assisted both Boyle and Newton.

Above: A modern electron microscope. In 1926 Hans Busch discovered that electrons could be used instead of light rays to achieve a much greater magnifying power. Parallel beams of electrons are made to converge by a magnetic field rather than a lens. The image is produced on a fluorescent screen. Modern electron microscopes can magnify specimens up to a million times and have made possible tremendous advances in research.

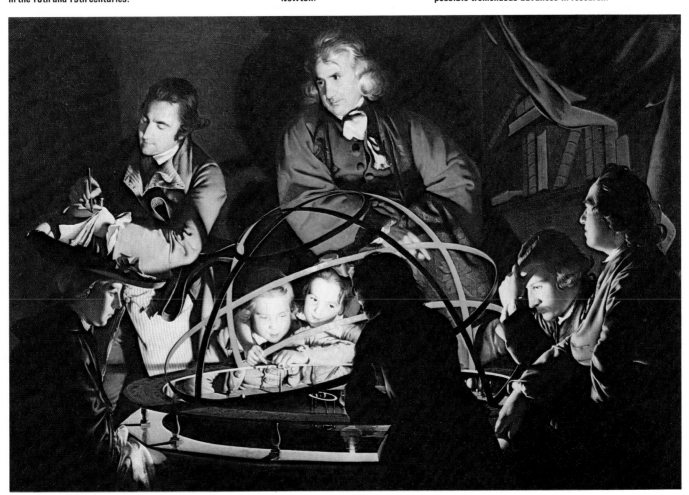

Reading the Rocks

Fossils in the rocks were long explained by the Biblical flood. Scientific study of extinct creatures, begun in the 18th century, soon became a fashionable pursuit.

Above: Mary Anning (1799–1847). Aged 11 she discovered the first ichthyosaur fossil and a few years later the first complete plesiosaur and the first pterodactyl in Britain.

Below: 'Dinner is served' . . . in the Iguanodon. A stage in the reconstruction of a dinosaur, in 1854, for the Crystal Palace.

In 1854 the Crystal Palace, home of the Great Exhibition of 1851, was moved from its original site in Hyde Park, London, to Sydenham in south London. Victorian sightseers were then treated to a display of lifesize, three-dimensional models of dinosaurs, ichthyosaurs, and other extinct animals, placed among the trees and shrubbery of the Crystal Palace grounds. The study and collecting of fossils had become a 'fashionable science', and one in which almost anyone could participate.

Before the Flood

Since before the time of the ancient Greeks, men had known about fossils and observed their similarity to some living sea creatures. In Renaissance times the accepted explanation was that Noah's flood had covered the continents, and that when the waters had receded, these creatures had been stranded and died. It was a convenient explanation because it agreed with what was written in the Bible. Even as late as the 18th century, a Swiss naturalist, Johann Scheuchzer, described fragments of fossil bones that had

been found in Germany as being those of men that had existed before the Flood. They were later identified as belonging to a giant salamander.

Two men, both born in 1769, can be credited as the first to read the world's history from its rocks: the French naturalist Baron Georges Cuvier, and William Smith, an English geologist. Cuvier, the founder of comparative anatomy, applied to extinct creatures his methods of classification of living animals by means of anatomical details rather than external appearances. This made it easier to relate extinct animals to one another as well as to their modern descendants.

While Cuvier enjoyed the best of educations, William Smith, son of an Oxfordshire blacksmith, was entirely self taught. None the less he made the first geological maps of England and Wales, and pointed out that

Below: A nest of dinosaur eggs, belonging to a Protoceratops.
Bottom: Edward Cope (1840–1897), a great fossil hunter whose spectacular finds helped to make paleontology a popular science.

particular fossils always occur in the same layers of the Earth's crust. For the first time these layers could be dated, since scientists knew when the fossilized animals in them had lived. Smith had provided later workers with what has been called 'the key to all geological research'.

In the 19th century the infant science of paleontology – the science of the study of fossils – grew in popularity. This was largely due to the work of men like the English paleontologist Gideon Mantell, and Richard Owen, an English anatomist. Mantell found the remains of the dinosaur *Iguanodon*, a 15-foot-high reptile, in Sussex, England in 1822. Richard Owen analysed *Iguanodon*, *Megalosaurus*, and *Hylaeosaurus*, and concluded that they were not the ancestors of modern lizards, but belonged to an extinct sub-order, which he named *Dinosauria* (from the Greek *deinos*, meaning terrible, and *sauros*, meaning lizard). Owen was in charge of the building of the great models for the Crystal Palace, constructed under his direction by the sculptor Waterhouse Hawkins.

American finds
The really spectacular fossil finds of the 19th century came from central and western America. In 1871 Edward Cope, an American naturalist, set out to find and describe all the extinct fishes, reptiles, and mammals of the western USA, from Texas to Wyoming. At times, threatened by hostile Indians, he had to collect his fossils with a rifle in one hand. Cope was one of the first to realize that the really important work begins after the fossil has been found, when it must be identified and classified.

Cope's great rival was Othniel Marsh, a paleontologist. A race developed between the two men to find the first fossil dinosaurs. Marsh was aided in his searches by the financial backing of his uncle, George Peabody, who built the Peabody Museum of Natural History at Yale University to house Marsh's finds. The enmity between Marsh and Cope had its advantages. Between them they managed to find and identify more than 130 species of dinosaurs, including *Brontosaurus*, *Diplodocus*, and *Hadrosaurus*. A huge mass of specimens was accumulated.

Cuvier had stated categorically that *'L'homme fossile n'existe pas'* ('Fossil Man does not exist'). This was very far from the truth, but the earliest fossil remains of men were not recognized as such. Only after Darwin and his followers had gained acceptance for the idea that Man was descended from ape-like creatures did the search for the fossil remains of his ancestors become widespread. The paleontologists travelled to Java, Peking, and Africa to piece together the history of Man's evolution.

The early paleontologists and fossil hunters built up an increasingly detailed picture of early life on Earth. The search continues for the vital clues to the living world thousands of years before the first men appeared on Earth.

Above: The skeleton of a dinosaur under reconstruction. There were often only isolated bones to work from and the overall shape of the dinosaur had to be guessed at.

Above: Othniel Marsh (1831–1899). His bitter race against Edward Cope to discover the first fossil dinosaurs led to the discovery of so many specimens that his uncle built the Peabody Museum at Yale to house them.

A New Form of Energy

William Gilbert finally disproved the belief that magnetism and electricity were forms of magic. His work raised them to a level worthy of serious scientific study.

Electricity and magnetism have been known to Man, in their natural forms, for thousands of years. In about 600 BC the Greeks found that amber attracted feathers and other light objects when rubbed vigorously. Lodestone, a naturally occurring ore, was found to attract pieces of iron, and compasses were probably used as early as AD 700. But such properties were generally regarded as magical. Lodestone was said to have many medicinal benefits, and the Romans believed that sufferers from gout would be healed by the shocks received from 'electric fishes'.

William Gilbert (1540–1603), an English physicist and court physician to Elizabeth I,

took a more scientific approach. He studied magnets and compasses and put forward a theory, published in 1600 in his book *De Magnete*, that the Earth is a huge magnet, and that this explains the direction of magnetic north and the dipping of the magnetic needle in a compass.

First electric machines

Electricity and magnetism from now on were regarded as natural, not magical. In 1672 Otto von Guericke (1602–1686), mayor of Magdeburg, constructed a machine which could produce static electricity in greater quantities than had been possible before. His machine was simply a ball of sulphur which rotated at great speed. When Guericke rubbed his hand against the spinning ball, an electric charge was induced in it, and he saw that it would attract light objects and then repel them after contact. At greater speeds the ball sparked. This was the first clue to the similarity between static electricity and lightning. It was another 80 years

Above: In 1752 Benjamin Franklin demonstrated the electrical nature of lightning by flying a kite in a thunderstorm and drawing sparks from a key tied to the lower end of the string.

The Leyden jar was invented in 1746 as a means of storing electricity. The jar is half-coated with tinfoil and is corked at the top. A brass rod passes through the cork and down to the foil at the bottom. Charges are passed down the rod to the tinfoil in which they are stored.

Above: Humphry Davy (1778–1829), a brilliant English chemist. He is chiefly remembered for inventing the miner's safety lamp, named after him, but he made many other contributions to 19th-century science. By inhaling nitrous oxide he discovered its anaesthetic properties, but nearly killed himself when he performed a similar experiment with methane. He was also the first to isolate the elements sodium, potassium, barium, magnesium, and strontium. In the light of these achievements he hardly merits the declaration sometimes made that his pupil, Michael Faraday, was the greatest of his discoveries.

Below: Otto von Guericke's static electricity generator, invented in 1672. Guericke's other main contribution to science was the invention of the air pump in 1650.

before this link was established by the American statesman, philosopher, and scientist, Benjamin Franklin (1706–1790).

Franklin's kite

Franklin believed that lightning was a discharge of electricity from the atmosphere to the Earth. To test his theory he performed a dangerous experiment. During a thunderstorm he flew a kite with a metal key on the end of a string. At the tip of the kite was a metal spike. Electricity passed from the atmosphere into the spike and down the wet string to the key where a spark appeared. Franklin had proved the existence of electricity in the cloud. This experiment led Franklin to invent the lightning conductor which would 'secure houses, churches, ships from lightning . . . by drawing the electrical fire out of a cloud silently, before it had time to strike.'

Further progress in electrical science was hampered by the lack of a steady supply of electricity. The Leyden jar, invented in 1746 in Leyden, Holland, could store static electricity. Franklin gave the jar its characteristic form, coating it with tin foil. Using Franklin's jars, William Watson sent a current through two miles of wire. But the Leyden jar could not produce a steady current.

The voltaic pile

An important breakthrough was made in 1800, when Alessandro Volta (1745–1827), an Italian physicist, constructed the first electric battery. The basic unit of the 'voltaic pile' was a piece of zinc and a piece of copper separated by a fabric pad soaked in a dilute acid. This produced a current, flowing from one metal to the other, and by building a 'pile' of these units, he could intensify the flow of current. Scientists now had a manageable source of current which they could use in their experiments. Using the voltaic pile, two English experimenters, W. Nicholson and A. Carlisle, showed that passing a steady current through a solution of salt and water resolved the water into its two parts, hydrogen and oxygen.

Another experimenter of the time who had great success in his work with Volta's pile was Humphry Davy (1778–1829). One of his most important experiments was the decomposition of molten soda, previously thought to be an element, into the metal sodium and water, by passing an electric current through it – a process known as electrolysis. A similar experiment yielded the element potassium.

Davy was a brilliant chemist and famous as an eloquent lecturer. He was appointed Professor of Chemistry at the Royal Institution, London, when he was only 24. Although best remembered as the inventor of the miner's safety lamp and the discoverer of the anaesthetic properties of nitrous oxide (laughing gas), his greatest contribution to science was his work on the electrical nature of chemical bonds – work that was to be carried on by his assistant, Michael Faraday.

ANIMAL ELECTRICITY

Luigi Galvani (1737–1798), professor of anatomy at Bologna, spent 11 years studying the effect of electricity on the nerves and muscles of animals. Many of his experiments involved passing electric currents through frogs' legs, which would twitch as the muscle contracted. In 1786 he noticed that if the nerve in a pair of legs was connected to a copper rod and the muscle to an iron one, and the two metals were also in contact, the legs would twitch although he had not applied any external current.

Galvani concluded that 'animal electricity' was being produced by the tissue. He was mistaken; the electric current was produced by the chemical action of very weak acids in the legs on the two metals, which being in contact, formed a complete circuit. But Volta used Galvani's discovery in inventing the first battery, and his name lives on in words like 'galvanometer', an instrument for measuring electricity, and the metaphor 'to be galvanized into action'.

MICHAEL FARADAY 1791–1867, the son of a Surrey blacksmith, laid the basis of electro-chemistry and of modern atomic and electronic sciences.

JAMES CLERK MAXWELL 1831–1879 used a mathematical description to support Faraday's discovery of electro-magnetism. He showed that light itself is electromagnetic in origin and predicted the existence of other electromagnetic waves. He also did valuable work on the theory of colour and gases in motion.

HERMANN HELMHOLTZ 1821–1894, a German physicist, is noted for his contributions to optics, acoustics, and electromagnetism.

HEINRICH HERTZ 1857–1894 was born in Hamburg and trained first as an engineer. As a physicist, he is famous for his experiment with Hertzian or electromagnetic waves. He showed they had the same properties as light and heat waves.

Towards Wireless Waves

Michael Faraday demonstrated the link between magnetism and electricity. The way was open for other scientists to put electricity to work.

The work of Faraday and others in the 19th century led to a clearer understanding of the link between electricity and magnetism. This was to be a bridge to new and more sweeping discoveries.

In 1820 the Danish physicist and chemist, Hans Christian Oersted, noticed, while lecturing in Paris, that when he passed an electric current through a conductor it made a nearby magnetic needle deflect. He had stumbled across the link between electricity and magnetism. Within a few weeks the French scientist André-Marie Ampère had shown that when a current is passed through a wire it produces a magnetic field. He also showed that if currents are passed through parallel wires, depending on the directions of the currents the wires will either attract or repel each other. Wires carrying currents clearly behaved like magnets.

Generating electricity

Oersted and Ampère had shown that a magnetic field could be created by an electric current. Michael Faraday, an English chemist and physicist, was to show that the reverse was possible too. Faraday was born in London in 1791. At the age of 12 he took a job as an errand boy in a bookshop. He had time to read many books, and became interested in the new science of electricity. In 1812 he sent the eminent chemist, Sir Humphry Davy, notes of Davy's own lectures at the Royal Institution. Davy was impressed and offered Faraday a job as his laboratory assistant.

In 1831 Faraday performed two very significant experiments in the field of electromagnetism. In the first, he showed that an alternating current in one conductor will induce a current in a second nearby conductor, although there is no electrical connection between them. In the second, he demonstrated that when a magnet is moved in and out of a coil of wire, a current is created in the coil. Faraday had discovered two ways of generating electricity. The dynamos, transformers, and electric motors later invented were a direct result of his discoveries. Without them there would be no way of producing electricity in large

One of Michael Faraday's most important feats was to establish the laws of electrolysis – a process of bringing about a chemical reaction by passing an electric current through a solution. One practical application of electrolysis is in electroplating. Below, a spoon is being coated with chromium from a solution of chromic acid through which a current is passed.

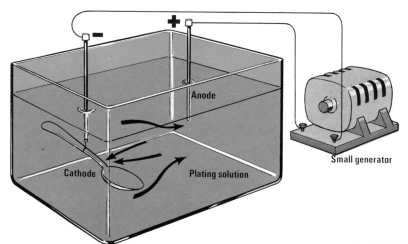

Anode

Cathode

Plating solution

Small generator

quantities or efficient means of distributing it.

Following this Faraday worked on the theory of electrolysis. He showed that a precise amount of electricity is needed to free a precise amount of an element from an electrolyte (the substance in solution), and that there must be a minimum unit of electricity. Faraday's electrical unit was, in fact, the charge carried by an electron.

Electromagnetic waves

Much had been discovered about electricity and magnetism, but the relationship had still not been mathematically expressed. Nearly 30 years after Faraday's discoveries James Clerk Maxwell (1831–1879), a Scottish physicist, used Faraday's experimental results to produce equations which showed the precise relationship between electricity and magnetism. He believed that a vibrating electric charge would produce electromagnetic waves (waves which have both electric and magnetic properties). He theoretically calculated their velocity and saw that it was the same as the speed of light. From this he correctly concluded that light itself was a form of electromagnetic wave. Maxwell only guessed at the existence of these waves. A German physicist, Heinrich Hertz (1857–1894), was to prove it.

Hertz initially trained as an engineer, but he became interested in physics and learned much as the student of another great German scientist, Hermann Helmholtz. Helmholtz's own researches covered numerous sciences from physiology to mechanics. During his work on electromagnetism he realized the need for a test of Maxwell's theory, and he made it the subject of a competition for students. His brilliant pupil Hertz successfully performed the test in 1888 and proved the existence of electromagnetic (or Hertzian) waves. Hertz found that they had identical properties to those of light and heat waves, except for a lower frequency. It was becoming clear that there was a whole spectrum of electromagnetic waves – the visible spectrum of light, and the invisible spectrum ranging from gamma rays to the ultra-violet, and infra-red to radio waves. Hertz was, in fact, very close to discovering X-rays, another form of electromagnetic wave, when he died at the age of 37, in 1894. His work had conclusively established the existence and form of electromagnetic waves, and prepared the ground for the development of radio.

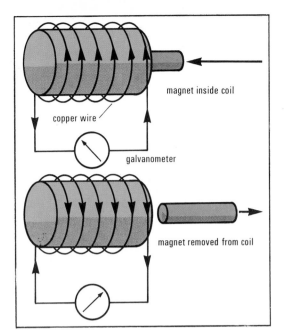

In 1831 Faraday demonstrated that when a magnet is moved in and out of a coil of wire a current is created in the coil. The galvanometer registering the current shows it flowing one way when the magnet is moved in, and the other way when the magnet is moved out.

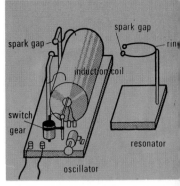

Above: Hertz's oscillator and resonator, which confirmed the existence of electromagnetic waves. A current produced in the induction coil causes a spark to jump across the gap between the terminals. Electromagnetic waves thus released set up a similar current and spark in the copper ring of the resonator.

Left: Two of the machines invented by James Wimshurst (1832–1903) for generating static electricity. When the handle is turned, the machine builds up a charge in much the same way as a comb creates electricity when it is drawn rapidly through the hair.

The world's first generating station was opened by Thomas Edison in New York in September 1882. This new power source soon became widespread. The Deptford power station, London (below), was opened in 1889.

The Microbe Hunters

Through their work, Pasteur, Jenner, and Lister revolutionized medicine and gave Man a vital weapon in the war against disease.

The first 'vaccination' against smallpox was in 1796; by 1801 Edward Jenner reported that at least 100,000 people in Britain had been vaccinated. This cartoon by James Gillray mocks supposed effects of the cowpox germs.

In 1683 a Dutch draper, Anton van Leeuwenhoek, peered through a home-made microscope at a drop of canal water, and found it was seething with tiny creatures. He described these animalcules, as he called them, as being 1000 times smaller than the eye of a louse. He was the first to see the micro-organisms we call bacteria, and he sent an account of his findings with drawings of the bacteria to the Royal Society in London, of which he was a Fellow.

At that time it was commonly thought that the lower forms of life – including such things as weevils, protozoa, and even shellfish – could be produced by non-living matter. This process was known as spontaneous generation. The micro-organisms seen under the microscope also seemed to appear out of 'nothing' – Leeuwenhoek saw clear broths produce increasing numbers of them as days went by. Leeuwenhoek did not believe they came from spontaneous generation; 'We can now easily conceive,' he wrote, 'that . . . in all waters exposed to

Louis Pasteur proved the connection between micro-organisms and fermentation and disease. The inoculations he pioneered have saved the lives of countless millions of men and animals. His name has given us the word pasteurization for heat sterilization of such things as milk.

Pasteur inoculated patients with a weakened strain of a disease which prevented them from catching it thereafter. The forerunner of this practice was Edward Jenner (left), an English doctor who had immunized people against smallpox by inoculating them with the germs of cowpox, a similar but very much milder disease. Before this, smallpox was widespread, killing many people and leaving others hideously scarred. Earlier, people had tried a form of immunization with germs from people with mild cases of smallpox; but the resulting illness was not always mild, and sometimes even proved fatal.

the air, animalcules may be found; for they may be carried thither by the particles of dust blown about by the winds.' But it was two centuries before his theory was proved by the work of Louis Pasteur.

Louis Pasteur

Pasteur was born in 1822, the son of a tanner in south-east France. At first he showed a talent for painting and drawing, but at the age of 19 he decided to become a scientist and went to Paris to study chemistry and physics. His first important work was on

EDWARD JENNER

In the 18th century, smallpox was still an uncontrollable scourge. Thousands died in violent epidemics. Edward Jenner (1749–1823), a doctor in Gloucestershire, England, became interested in the local belief that cowpox (a disease of cattle which sometimes infects the hands of milkers) made people immune to smallpox.

In 1796 Jenner put matter from cowpox sores into a cut on the arm of James Phipps, an 8-year-old boy. When the resulting cowpox sores had cleared up, Jenner infected him with smallpox the same way. This should have produced a fatal attack; but the boy remained healthy.

Jenner's discovery was soon known throughout the medical world. Wide acceptance came surprisingly quickly. Jenner vaccinated thousands of poor people free of charge, and the number of smallpox cases dropped at once.

crystal structures; this led him on to wonder about the structures of living things and he began to investigate micro-organisms as the simplest forms of life.

In 1854 Pasteur became professor and dean of the faculty of science at Lille. At that time local manufacturers of alcohol were suffering appalling losses through the souring of alcohol during the fermentation process. Pasteur turned his attention to fermentation in wine and beer. He found that the sour liquids contained micro-organisms which prevented proper fermentation. He went on to show that micro-organisms also caused the souring of milk.

Where did these micro-organisms come from? Did they fall from the air, or were they spontaneously generated in the liquids? Ever since Leeuwenhoek's day scientists had fought

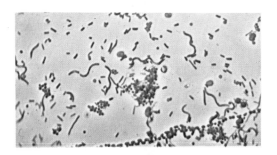

These bacteria (magnified × 740) fell from the air on to a coated plate. They are of different shapes and sizes; some are harmless, others can cause disease. Pasteur was the first to realize the effects of these invisible airborne hordes.

for these rival theories. Now Pasteur, in a series of complicated experiments, showed conclusively that while a sterilized liquid protected from atmospheric dust would remain pure for months, micro-organisms were growing in it only a few hours after it was exposed to the open air. The supporters of spontaneous generation were finally vanquished.

Pasteur showed that not only fermentation but also the decomposition of organic matter was caused by certain micro-organisms, or 'germs', in the air. He showed

In the 18th century surgical techniques were necessarily crude; operations had to be carried out as fast as possible since there were no anaesthetics. By the mid-19th century techniques of surgery were greatly improved, but the death rate from wound infection was terrifyingly high.

Joseph Lister (1827–1912), influenced by Pasteur's work, revolutionized medicine by preventing the contamination of wounds by micro-organisms in the air; he founded antiseptic surgery.

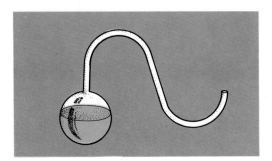

Left: With this flask Pasteur proved that micro-organisms are not generated from dead matter. The broth inside, boiled to kill any living things, did not decay, because the microbes in the air were trapped in the bent neck of the flask. But when the neck was snapped off the broth began to decay in a few hours.

how these could be killed by heat, in the process called pasteurization named after him. His work came to the attention of a British surgeon called Joseph Lister, who realized that germs might be the cause of the wound infections which made surgical operations so dangerous. Lister could not use heat to kill the germs; he killed them instead by placing an antiseptic 'barrier' of carbolic acid between the wound and the air. Deaths from wound infections dropped dramatically. Other scientists realized that germs could be carried by the hands, and by instruments. Over the years they fought to spread their ideas, sometimes against fierce opposition; until eventually the killing of germs by antiseptics and heat sterilization became common practice.

In 1868 Pasteur was partially paralysed by a stroke, but this did not halt him. For he believed that: 'Will, work, and success between them fill human existence.' He went on to isolate the bacteria that cause chicken cholera and anthrax in sheep and cattle. He found that animals injected with specially weakened strains of the bacteria suffered from only mild attacks of the diseases – and, far more important, they later proved immune to full-strength bacteria. Pasteur's last great success was in producing a vaccine to immunize people against the fatal and agonizing disease rabies, which they contracted by being bitten by an infected animal.

Pasteur's work, and that of his followers, proved of vital importance in our understanding of diseases and their cures.

Linking the Continents

Samuel Morse (1791–1872), inventor of the electric telegraph. He also devised the Morse Code.

Scientists had proved the existence of electromagnetic waves which could be transmitted through the air without wires. Now they searched for a way to use these waves to transmit information — and then speech.

Alexander Graham Bell, inventor of the telephone, inaugurating the New York – Chicago line in October 1892.

Electricity had been discovered. Now scientists found it presented new possibilities in the important realm of communications.

The first clue that electricity might be the key to rapid communications was provided in 1820 when Hans Oersted, a Danish physicist and chemist, discovered that an electric current could deflect a magnetic needle. The importance of his discovery was quickly appreciated by the German mathematician, Karl Gauss (1777–1855). He used different strengths of current to indicate different letters of the alphabet, so that as each letter was transmitted, a magnetic needle was deflected to a different position.

Messages by wire

From this idea came the first electric telegraph, built and demonstrated in 1837 by Samuel Morse (1791–1872), an American artist who became interested in electricity after hearing of experiments in electromagnetism during a trans-Atlantic voyage. A key at the sending end was used to produce long and short bursts of current – dots and dashes – in what became known as the Morse code. At the receiving end a pen was moved by an electromagnet to inscribe these dots and dashes on a moving band of paper. Morse had to fight to get his telegraph system accepted. In 1849, angry farmers tore down his telegraph wires because, they said, the electric currents upset the weather and spoiled their crops. By the time Morse died, messages in Morse code were being carried to all parts of the world.

The chief drawback to the use of the Morse telegraph was that its operators had to be skilled in the use of his code. The ideal, the transmission of speech, seemed an impossibility until the work of Alexander Graham Bell (1847–1922), a Scottish teacher settled in America, proved otherwise.

'Mr Watson, come here . . .!'

Bell did not know much about electricity, but he understood the human voice, for he was professor of voice physiology at Boston University. He believed it was possible to transmit all the sounds of the human voice over a telegraph system. In June 1875, while experimenting with the help of his assistant, Thomas Watson, Bell heard a noise over his machine from the room where Watson was working. It was the first step towards a working telephone. In March 1876 the first intelligible sentence was transmitted over the telephone. Bell and Watson were about to try out a new transmitter when Bell cried: 'Mr Watson, come here. I want you!' The message was real and urgent, for Bell had spilt some battery acid down his trousers. Watson, hearing Bell's cry on the receiver in the next room, rushed in triumph to Bell. Their invention was a success.

The next step forward came with wireless communication. In 1888 the German physicist Heinrich Hertz became the first to demonstrate the production and reception of electromagnetic waves. He measured the speed and length of these waves and showed that although they travelled at the same speed as light their wavelength was very much longer. When Hertz discovered these waves, he thought that they would be of theoretical interest only, but in fact they paved the way for radio.

In 1890 the French physicist Edouard Branly invented a receiver for electro-

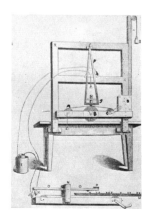

The original Morse telegraph. Messages are sent by means of a Morse key and a sounder, connected to a low-voltage electric circuit. The sender makes long and short electric pulses with the key, which cause the sounder (an electromagnet) to click against an iron bar.

magnetic (or Hertzian) waves, which he called a coherer, and four years later a Russian, Aleksander Popov, invented an aerial which improved the range over which electro-magnetic waves could be transmitted. The groundwork had now been done, and the way was open for the young Italian scientist, Guglielmo Marconi (1874–1937), to produce the first 'wireless'.

The first radio message

In 1894 Marconi used Hertzian waves and a coherer to make a bell ring at a distance of 15 feet from the source. His father was not impressed. 'After all, there are other ways of making a bell ring,' he said. None the less Marconi developed his transmitter, and by 1899 he had increased its range to 75 miles. Scientists thought that there would be a limit to the distance that these radio messages could be sent. After all, radio waves travel in straight lines, and the Earth's surface is curved. At some point, they thought, the signal would be lost in space. But in 1901, Marconi transmitted the Morse letter 'S' from England to Newfoundland, a distance of 5000 miles. The radio waves had been bent round the Earth by bouncing them off

a layer in the upper atmosphere, called the ionosphere. Radio was now a reality.

From Marconi's first radio, our systems of communication have rapidly expanded. Television, and the ability of astronauts to talk to us from the Moon, now seem merely a matter of course. Radio astronomers receive signals from millions of light years away in space. The scope for communication today appears limitless.

Above: The inventor of radio, Guglielmo Marconi, with apparatus similar to that which he used to send the first wireless signal across the Atlantic – the Morse letter 'S', transmitted from England to Newfoundland in 1901.

TELEVISION

The discovery of radio led, some 25 years later, to the development of television – a more complex instrument but based on the same principles. Sound, for both radio and television, is converted into electrical signals which are strengthened and then combined with a radio wave which carries them through the air. For television, pictures also have to be converted into electrical signals. Below is a simple diagram which shows how colour transmission is achieved.

The light from the scene to be televised is split by mirrors into the three primary colours — blue, green, and red. The light rays are converted into electrical signals by three image orthicons, and the signals are strengthened by an encoder before transmission. The adder forms black and white signals. The signals

are transmitted and sorted out into the three colours by the television receiver. Electron guns in the cathode tubes fire scanning beams on to the screen, which is coated with three kinds of phosphor dots which glow respectively blue, green, and red when struck by the beam.

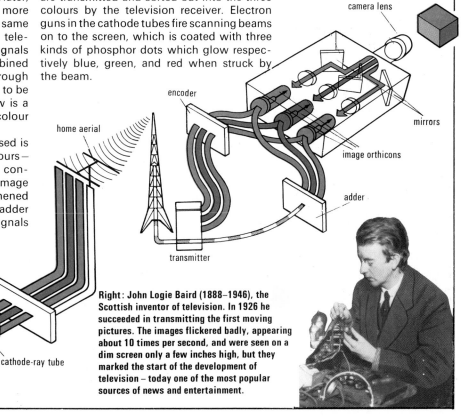

camera lens

encoder

home aerial

mirrors

image orthicons

adder

transmitter

colour TV receiver

cathode-ray tube

Right: John Logie Baird (1888–1946), the Scottish inventor of television. In 1926 he succeeded in transmitting the first moving pictures. The images flickered badly, appearing about 10 times per second, and were seen on a dim screen only a few inches high, but they marked the start of the development of television – today one of the most popular sources of news and entertainment.

Exploring the Spectrum

Until the mid-19th century, men had studied only the visible part of the spectrum — light waves. The accidental discoveries of X-rays and natural radioactivity proved that light was only one small part of a much wider band of electromagnetic waves.

Isaac Newton had shown that white light can be broken up by a prism into a spectrum of colours. By the beginning of the 19th century it was recognized that this visible spectrum was only a small segment of a more extensive and invisible spectrum that we call the electromagnetic spectrum. This spectrum is composed of a whole family of radiant waves ranging from radio waves to X-rays and gamma rays. The true nature of the electromagnetic spectrum emerged from research into radio transmission but was aided by the discoveries of the pioneers of photography.

Capturing light

The ancient Greeks were familiar with the 'camera obscura' or pinhole camera, which could produce an image on a translucent screen. Yet nobody had ever managed to capture these elusive images in a permanent

way. The first person to do so successfully was the French inventor Louis Daguerre (1787–1851). Daguerre stumbled upon his discovery in 1839 after he had been exposing in a camera a silvered plate coated with silver iodide in the hope of obtaining an image. The experiment failed and he put the plate away in a cupboard containing chemicals. Later he discovered that a brilliant image had appeared on the plate and that this was due to vapour that had escaped from a bottle of mercury in the cupboard.

In England William Fox Talbot (1800–1877) developed another process of photography in 1841 that became the starting point of modern photography. Talbot's 'calotype' process made it possible to make as many positive prints from the original negative as might be needed. Photography became an important new tool for scientists.

The equipment of a 19th-century photographer was staggeringly clumsy compared with today's easily portable cameras. He trundled his camera round on a barrow, shrouding himself with the hood to take a photograph and to develop the wet-coated plates which were used until the late 19th century.

Human eyes are sensitive only to visible light. They are unable to see radio and television signals or X-rays. Yet visible light rays are only a very small segment of the entire electromagnetic spectrum. Electromagnetic rays travel as waves and vary in length from 18-mile-long radio waves to microscopic gamma rays. They travel at the speed of light – 186,000 miles a second.

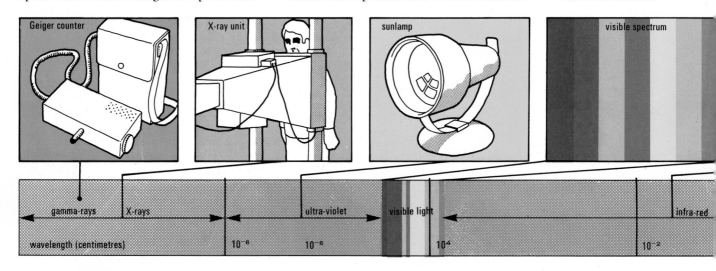

Geiger counter   X-ray unit   sunlamp   visible spectrum

gamma-rays   X-rays   ultra-violet   visible light   infra-red

wavelength (centimetres) 10^{-8} 10^{-6} 10^{-4} 10^{-2}

Pictures of the invisible

In 1895 Wilhelm Roentgen (1845–1923), a German professor of physics, was researching the properties of cathode rays. While working with a Crookes' tube, a glass vacuum tube through which an electric current can be passed, Roentgen found that some nearby photographic plates had become fogged in spite of being protected inside lightproof wrappings. The unknown emission from the Crookes' tube also caused a fluorescent screen to glow, even when the tube had been made lightproof with cardboard. More astonishing were the eerie shadow pictures of the bones of Roentgen's hand that the rays cast when he held it between the tube and the screen. Unable to explain the puzzling penetrating power of these rays or why they could only be detected by their effects on photographic plates, Roentgen named them X-rays, X being the symbol for the unknown.

A similar chance discovery was made by the French mineralogist, Antoine Henri Becquerel (1852–1908). Becquerel first found that minerals containing uranium, after exposure to sunlight, had the same effect on wrapped photographic plates as X-rays. More amazing was the accidental discovery he made a few days later. Becquerel happened to be developing some photographic plates which he had placed in a drawer in which there were traces of a uranium compound. He was startled to see that they had been blackened by the uranium. Becquerel had discovered the existence of natural radioactivity. He also found that uranium ore was far more intensely radioactive than uranium itself; it was this discovery that especially puzzled Marie Curie (1867–1934) and her husband Pierre.

A new radioactive element

The Curies set about isolating the other mineral in uranium ore that they believed was the source of the additional radiation. From the Austrian government they obtained several tons of free pitchblende residues — uranium-bearing ore from which the uranium had been extracted. They began a laborious process of 'cooking' the pitchblende to extract the unknown mineral and after two years produced a new radioactive element that Marie named 'polonium' after her native Poland. In December 1898, they isolated a tiny quantity of a second element they called 'radium'. Here was a mineral that, in its pure form, proved to be a million times more radioactive than an equivalent amount of uranium.

The discovery of radioactivity and the isolation of radium was the turning point in modern physics that led to the investigation of the structure of the atom and the relationship between matter and energy.

Marie and Pierre Curie in their laboratory in 1903 – the year of their joint Nobel Prize. Pierre died in 1906 but Marie won a second Nobel Prize for work on radium in 1911.

Antoine Henri Becquerel (1852–1908) was the discoverer of natural radiation and shared the Nobel Prize with the Curies in 1903.

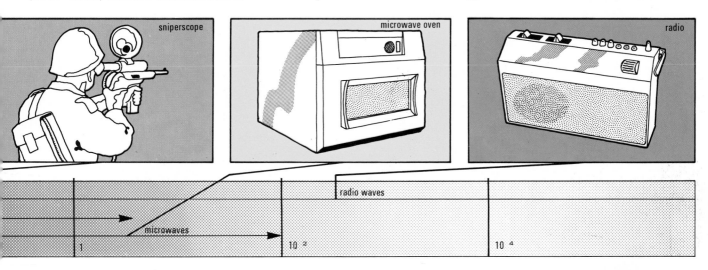

The Nature of Energy and Matter

James Clerk Maxwell (1831–1879) temporarily overthrew Newton's 'corpuscular' theory of the movement of light. He sent his first paper, on the 'Description of Oval Curves', to the Royal Society of Edinburgh when he was only 15.

Max Planck (1858–1947), the German physicist, proposed and developed the quantum theory of radiation. He was awarded the 1918 Nobel Prize for physics.

Below: Lord Rutherford holding his original alpha-particle scattering apparatus.

Newton's laws governing the universe were modified in the 20th century by Albert Einstein — a scientist who worked not in a laboratory, but with pencil and paper.

The elegant laws of the physical universe that were expounded by Newton in the latter half of the 17th century remained essentially unchallenged for the next two hundred years. But in the middle of the 19th century, the brilliant Scottish scientist James Clerk Maxwell postulated that light, and every other form of radiant energy, travels in the form of electromagnetic waves which radiate outward from their source much like the ripples that are formed by dropping a stone into a pool. Newton's disputed 'corpuscular' theory of light, which maintained that light consists of tiny particles, was at last disproved — at least so it seemed.

Essential to Maxwell's wave theories was the concept of an invisible 'ether' that permeated the universe. Light waves were believed to be transmitted through the ether much as sound waves are transmitted by molecules in the air. But in 1887, two American scientists, Albert Michelson (1852–1931) and Edward Morley (1838–1923), showed that the speed of light never varies and that Clerk Maxwell's ether does not exist. Their findings were later used by Einstein in developing his famous 'theory of relativity'.

As the 19th century drew to a close, Max Planck (1858–1947) in Berlin showed that radiation, such as light and heat, is given off in the form of tiny 'packets', or particles, of energy, as well as in the form of waves. Planck called these particles quanta, but he was somewhat reluctant to believe his own theory as it disagreed with the former idea that radiation was a continuous stream of energy with a wavelike motion. Although the complete implications of the quantum theory were not realized until the 1920s, it was soon used by Einstein in his research into the nature of light.

Albert Einstein established the relationship between mass and energy with his equation $E = mc^2$, where E is energy, m is mass, and c the velocity of light. Einstein was born in Germany and worked in Switzerland and Germany before becoming a US citizen when his Jewish origins made life in Germany intolerable under the Nazi regime.

Einstein and the revolution in physics

If it can be said that modern science has a man of genius, that man is undoubtedly Albert Einstein (1879–1955). In spite of the prophecy of one of his school teachers that he would never amount to anything in life, Einstein has been responsible for re-ordering mankind's view of the universe. In 1905, he wrote a series of scientific articles that boldly swept aside the long established 'facts' of Newtonian physics.

Einstein was already familiar with Planck's quantum theory when he proposed that light can also exist as bundles of energy, or photons. However, Einstein's findings did not imply a rejection of Maxwell's wave theories. Rather, they showed that radiated energy was a two-faced phenomenon. Thus the nature of low-frequency radiation (radio waves) is best described in terms of waves, whereas photons are more suitable to describe high-frequency gamma and X-rays. Visible light possesses both wave and particle characteristics.

$E = mc^2$

On the Electrodynamics of Bodies in Movement was the 1905 article in which Einstein outlined his famous theory of relativity. He rejected the notion that absolute measurements of movement can be made in a universe which lacks any single fixed point. All motion is therefore relative to the place at which it is measured. The only constant 'fact' in the universe is the speed of light: 186,272 miles per second.

The theory of relativity also introduced the famous $E = mc^2$ equation that established the link between energy and matter. Traditional physics had been anchored to the notion that a fixed amount of matter existed in the universe which could neither be destroyed nor added to. But Einstein's formula proved that matter and energy are two aspects of a single phenomenon. It also showed that a tiny bit of matter could be converted into a stupendous amount of energy. One pound of matter, if totally destroyed, could theoretically yield more energy than 10 million tons of ordinary explosive. Einstein himself was cautious about the implications of this formula at first. 'There is not the slightest indication that the energy will ever be obtainable. It would mean that the atom would have to be shattered at will. . . .' Yet only 40 years later, the fireball of the first atomic bomb became a monument to the determination of that will.

Rutherford's invisible planetary system
The difficulty in investigating atoms is that they cannot be seen even under the most powerful microscopes. In 1896 the French scientist Henri Becquerel had discovered that uranium gives off a continuous stream of rays or radiation. Later the Curies isolated radium. At once scientists began to investigate the nature of these rays. Among them was a New Zealand-born scientist, Ernest Rutherford, who was Professor of Physics at McGill University in Canada. While at the university he established that the Curies' radium atoms were disintegrating all the time and were emitting three types of rays called alpha, beta, and gamma rays. From this discovery Rutherford put forward a revolutionary theory concerning the nature of matter and the atom's structure.

Rutherford's model of the atom was the result of a famous experiment during which he bombarded fine sheets of gold foil with a stream of alpha particles. Most of these relatively large particles penetrated the foil, indicating that atoms are largely empty space. Yet a small number of alpha particles were inexplicably deflected – rather as if a

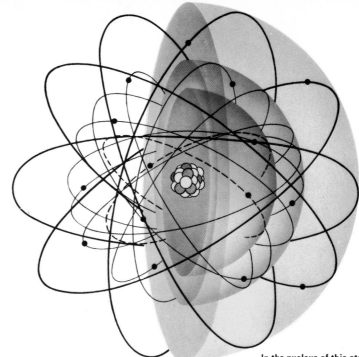

warship's shell were to bounce back after striking a sheet of tissue paper. Rutherford's findings contradicted the then accepted theory that atoms were solid spherical masses in which electrons were embedded like raisins in a pudding.

In 1911 Rutherford announced a new model of the atom based on his experiments. It showed that virtually all of an atom's mass is concentrated in a tiny nucleus consisting of positively charged particles known as protons. (In Rutherford's earlier experiment the few alpha particles that bounced back were those few that managed to hit the nucleus.) Around the nucleus, negatively charged electrons whirl in orbit like the planets round the Sun.

In the nucleus of this atom of phosphorus are 15 positively charged protons and an equal number of neutrons, which have no charge. Round the nucleus orbit 15 negatively charged electrons. It was Rutherford who first described the orbiting of the electrons around the nucleus – though at that time the existence of neutrons and protons was unknown.

Below: A characteristic mushroom cloud marks the spot of an atomic explosion. The devastating force of an atomic bomb is caused by the release of all the energy in an atom in a fraction of a second. Einstein had described the power locked within the atom – but had stated that to split the atom and release the power was impossible.

Power from the Atom

The word atom means 'the indivisible thing'. But modern scientists have shown that this is very far from being an accurate description of the smallest particle of matter.

The birth certificate of the atomic age can be seen in a former squash court in the University of Chicago. It is a plaque reading: 'On December 2, 1942 Man achieved here the first self-sustaining chain reaction and thereby initiated the controlled release of nuclear energy.' The man who ushered in the new age was an Italian physicist called Enrico Fermi.

Bombarding the atom

The discovery of radioactivity and the work of J. J. Thomson, Ernest Rutherford, Niels Bohr, and others had shown that the atom is not an indivisible unit but consists of a nucleus made up of particles called protons and neutrons, surrounded by orbiting electrons. Rutherford had also shown that by bombarding the nucleus with the atomic particles emitted by radioactive substances one atom could be changed into another.

Above: Professor Niels Bohr (1885–1962), a great Danish physicist, operating a cyclotron (particle accelerator). Bohr was awarded the Nobel Prize in 1922 for his work on the structure of the hydrogen atom.

In 1934 Enrico Fermi began to bombard all the known elements with neutrons to see which would become radioactive. He found that his neutron 'bullets' were more effective if he slowed them down by passing them through a 'moderator' such as water. He was a keen sportsman, and to explain this he used

Below: An uncontrolled nuclear chain reaction. Neutron 'bullets' are fired at the element uranium-235, and absorbed. The uranium nucleus explodes with tremendous force, throwing out nuclei of lighter atoms like barium and krypton as well as more neutrons. These neutrons collide with other uranium nuclei, and so the process continues.

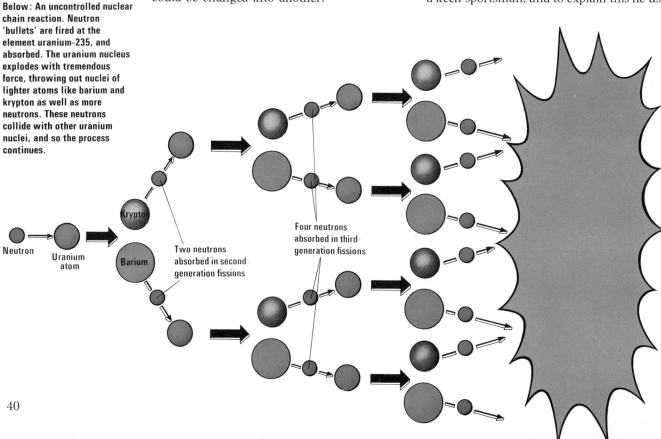

Neutron

Uranium atom

Krypton

Barium

Two neutrons absorbed in second-generation fissions

Four neutrons absorbed in third-generation fissions

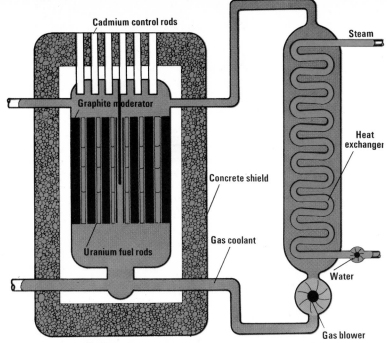

Cadmium control rods

Steam

Graphite moderator

Heat exchanger

Concrete shield

Gas coolant

Uranium fuel rods

Water

Gas blower

Nuclear reactor

The nuclear reactor contains a mass of uranium which gives off neutrons, and a graphite moderator to slow down the neutrons and facilitate the chain reaction. Cadmium rods are inserted or withdrawn to control the rate of reaction, and a liquid or gas 'coolant' is driven round the reactor to draw the heat off to a heat exchanger. Here, water is heated to produce steam to drive electric generators and turbines.
Below: Enrico Fermi (1901–1954), a great Italian physicist. When he produced the first nuclear chain reaction during the Second World War, he had to be careful how he told of his achievement. He sent a telegram reading, 'The Italian navigator has entered the new world'. The 'new world' was capable of the production of atomic bombs, which brought destruction on an unprecedented scale.

the analogy of a slow-moving golf ball, which is more likely to enter a hole than a fast-moving one. A slow neutron, said Fermi, was more likely to react with a nucleus with which it collided, and there was more time for the reaction to take place.

In 1938 Fermi received the Nobel Prize for his work on nuclear bombardment. He went to Stockholm in Sweden to receive his award – and from there sailed to the United States. For at that time Italy was under Fascist rule, and Fermi was opposed to its principles.

Controlled explosion

In 1939 the Austrian Lise Meitner showed that some of the atoms of uranium split into lighter fragments when bombarded with neutrons. The total mass of the fragments was less than that of the original atom; the missing mass must have been converted into energy, as Einstein had stated. When Fermi heard of this work he carried out experiments which indicated that the split uranium nucleus produced further neutrons; these could in turn split more uranium nuclei and, under the right conditions, the energy produced by this 'chain reaction' would produce an explosion of incredible magnitude.

How could this reaction be controlled? An answer lay in the speed at which the neutrons were travelling. Most of the neutrons released in the fission (splitting) moved so fast that they failed to split other atoms. To slow them down, Fermi embedded the uranium in a mass of graphite, which acted as a moderator. Graphite tends not to absorb

neutrons, but to reflect them back into the uranium, travelling at slower speed.

It was clearly vital that the chain reaction should not proceed too fast and produce an uncontrolled reaction. Fermi overcame this problem by inserting cadmium rods in his pile of graphite. These could be moved in or out of the uranium. Cadmium absorbs slow neutrons, so by moving the cadmium rods in and out Fermi could control the reaction.

In 1942 Fermi's pile was built at Chicago. It was in fact the world's first nuclear reactor. Although his work was then a step towards the atomic bomb, by showing how the chain reaction could be controlled he had pointed the way to the peaceful uses of atomic energy.

The New Cosmology

Above: The English astronomer
James Jeans (1877–1946), one
of the first advocates of a
'steady state' theory of the
creation of the universe.

Below: Dark patches, called
sunspots, frequently appear on
the Sun's surface at places
where the temperature is
abnormally low.
Below centre: Humason's
comet, an unusually large one,
seen in 1961. Although comets
have been observed for over two
thousand years, scientists still
do not know their composition.
The tail of a comet may be
over 100 million miles long.
Below right: The Rosette
Nebula – central stars
surrounded by gaseous
matter. Modern astronomy
has proved that while some
nebulae, like this one, are
within our galaxy, others are
in fact distant galaxies
themselves.

*The questioning of Newtonian
physics extended to astronomy
as advanced equipment revealed
new mysteries in the universe.*

Man's ideas of the universe have evolved
in step with the growth of his knowledge
and the sophistication of his observations
and measurements. Although the extent of
the known universe during Newton's life in
the 17th century was relatively limited because of the poor quality of telescopes at the
time, scientists had no great difficulty in
making sense of the phenomena they observed. Newton's laws of mechanics satisfactorily explained the movements of the
Sun, the Moon, and the planets in terms of the
gravitational forces acting upon them. But
as the horizons of the universe continued to
expand before the systematic searching of
astronomers, certain uncomfortably puzzling
phenomena began to emerge.

Charting the Milky Way
A leading figure in the exploration of the
frontiers of the universe was the great 18th-
century astronomer, William Herschel. His
painstaking improvements to the telescope
provided him with a research tool of un-
paralleled sophistication. Among his many
findings, Herschel's detailed observations of
the night sky resulted in the discovery of the
planet Uranus, thus doubling the known size
of the solar system, and in the charting of
more than 2300 nebulae. These mysterious
fuzzy patches of light were generally believed
to be distant luminous dustclouds, although
Herschel speculated that they might be

huge clusters of stars similar to the Milky
Way. From his observations, Herschel recognized that the Milky Way constitutes a distinct
celestial body in its own right and that our
solar system is no more than a tiny smudge
of light within this vast disc of stars.

The true immensity of the universe was not
fully appreciated until the early years of the
20th century. The accepted 19th-century
view of the universe was one which held
that our galaxy is unique. It is a Catherine
wheel of stars and star systems so vast that
light takes over 40,000 years to travel across
its diameter. There was great uncertainty
whether anything other than scattered dust
clouds lay outside the galaxy.

In the 1920s the American astronomer
Edwin Hubble, working with the enormous
100-inch telescope on Mount Wilson in
California, was able to discern that the light
of the nebula Andromeda originated from
the myriad pinpoints of individual stars.
Hubble's calculations showed Andromeda
to be some 800,000 light years away (a
considerable underestimate as it turned out),
more than eight times farther than the most
remote star in the Milky Way. He had shown
that our galaxy is not unique.

Using equations formulated by Einstein,
Hubble made the startling discovery that
Andromeda and all other galaxies are
hurtling away from one another at speeds
that become greater and greater as their
distances from one another increase. His
findings became known as 'Hubble's Law'
after they were published in 1929. He also
proved that the universe is steadily expanding
towards infinity in every direction, although
the stars and the galaxies within it all
remain the same size.

Big bang or steady state
Hubble's postulation of an expanding universe begged one obvious question – what

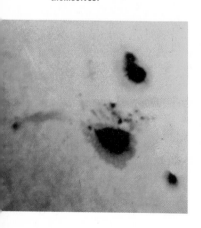

had caused it to expand in the first place? The Belgian priest, Abbé Georges Lemaître, attempted to resolve this problem. He argued that at some distant time in the past, all the matter in the heavens was crushed together in one enormous mass. This great cosmic egg finally exploded (the big bang) spewing out matter in every direction. The galaxies of today are nothing less than the remnants of Lemaître's original 'primeval atom'. Unfortunately, this theory does not explain how the matter was formed or why it all accumulated in one place.

The 'steady state' theory introduced a novel twist by asserting that matter is being continuously created throughout the universe. This idea was first suggested by James Jeans in 1928 and was championed by Thomas Gold and Fred Hoyle some 25 years later. The steady state theory proposes that matter spontaneously appears in the space where former galaxies, that have since moved away, were born. Thus the universe goes on recreating itself forever — out of nothing.

Recent findings in radioastronomy, including the discovery of quasars and cosmic radio signals which may in effect be the echo of an ancient 'big bang', have tended to undermine the 'steady state' hypothesis. Quasars, which are small galaxies, present a major problem for astronomers. They are the most distant objects in space, far more luminous than ordinary galaxies, and are receding at enormous speeds at the very fringes of the observable universe. Quasars may reveal what the universe looked like very early in its evolution.

A recent variation of the 'big bang' theory suggests that the universe is a rhythmically pulsating system. It regularly expands and contracts over and over again in a cycle that may last as long as 80 million years.

Hearing the universe

Since light is only a tiny segment of a virtually limitless spectrum, it is not surprising that it is not the only form of radiation emitted by the stars. With a telescope, astronomers have been able to see distant stars, whereas with a radio telescope it is also possible to hear them. Radioastronomy has the advantage of being able to penetrate dust clouds which obscure stars. The 'ear' of a radio telescope is a huge dish-like structure that can be aimed to eavesdrop on any part of the universe.

Above: The radio telescope at Jodrell Bank, England. Telescopes like this can 'listen' to and locate stars which emit radio waves, and are not always visible.

Below: Edward White making the first American space walk in 1965, a few months after that of the Russian, Alexei Leonov.

Below: From pictures like this of the Moon's surface and manned space flights to the Moon itself we have learned much about the Earth's only satellite, 240,000 miles away.

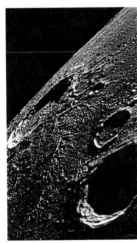

The Living Earth

The Earth under our feet feels absolutely solid and stable. But violent forces are confined beneath its surface, and even the continents themselves are slowly moving.

In 1650 Archbishop Ussher of Ireland calculated that the world had been formed on October 23, in the year 4004 BC, at 9 in the morning! Today it is thought that the Earth, the Sun, and the other planets in the Solar System were in fact formed about 4600 million years ago. That the world was far older than anyone had imagined was suggested by 19th century geologists – as a result of their systematic study of rocks and fossils. The question they could not answer was how the Earth and the Solar System came into being. It is a matter still debated by scientists today.

One theory suggests that a star passed very close to the Sun and drew off a huge cloud of gas which broke into fragments and cooled into the planets of the Solar System. Another theory put forward was that the Sun once had a companion star

A cut-away view of the Earth, revealing its structure. It is thought that the inner core may be of solid iron, under great pressure. The outer core is molten iron and nickel. The semi-liquid mantle is crossed by slow convection currents which cause volcanoes and earth movements. The 25-mile-thick crust is protected by a thin layer of atmosphere.

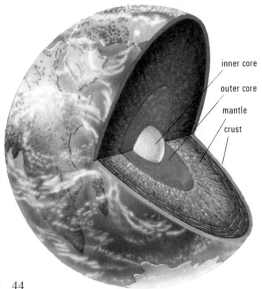

inner core
outer core
mantle
crust

Above: A volcano belches fire as a pool of magma within the Earth forces its way to the surface, erupting as red-hot lava.

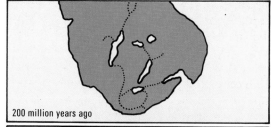

200 million years ago

100 million years ago

Two hundred million years ago the Earth's land masses were gathered together into the huge super-continent, Pangea. About 100 million years later movements beneath the Earth's surface split up Pangea. Australia and Antarctica broke free of Africa and the Atlantic Ocean began to take shape. Today the continents continue to move, riding on huge sections of the Earth's crust called plates.

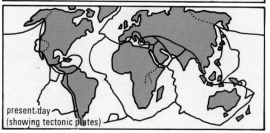

present day (showing tectonic plates)

44

Swirling clouds photographed from a satellite. Satellites orbiting Earth help meteorologists to study the weather.

An unusually well preserved fossil of an ichthyosaurus. Scientists have learned much about the Earth's history by studying the petrified remains of prehistoric life.

which exploded, scattering debris which became the orbiting planets. The most probable theory was first proposed in 1796 by the French astronomer and mathematician Pierre Laplace. His Nebular Hypothesis Theory maintained that there was a great cloud of dust and gas in the universe which became more and more concentrated under the influence of gravity. The centre of the cloud condensed into the Sun, and other centres of concentration became the planets. Modern research seems to confirm Laplace's cloud theory.

Inside the Earth

When the Earth was formed, the force of gravity pulled the densest matter into the centre to form the core. Less dense materials formed the mantle, and the least dense were forced to the outside and became the Earth's crust. Scientists have long believed that the core of the Earth is extremely hot and dense. The existence of the core was first established mathematically in the early 20th century. There is some evidence that the inner part of the core is solid, but this has never been proved.

In order to find out more about the internal structure of the Earth, scientists have bored holes into the crust and have taken rock samples. But even the deepest oil drillings only penetrate the crust, which is about 25 miles (40 km) thick under the continents. Below this crust tremendous pressures build up in the mantle – and from time to time break through the Earth's surface as volcanoes and earthquakes. When a volcano erupts, molten rock, or magma, from deep inside the Earth is forced through a point of weakness to the surface. Earthquakes are caused by the sudden readjustment of the rocks in the crust, which fracture and shift. Once a line

of weakness, or fault, has formed in the crust, the rocks are liable to go on shifting.

Our knowledge of the mantle and core of the Earth is gained entirely from recording and analysing the seismic (shock) waves produced by these disturbances. The first person to treat earthquakes as a scientific phenomenon was Robert Mallet, an Irish engineer, after he investigated a devastating earthquake in Naples in 1857. He later developed the technique of producing artificial underground explosions to measure the effects of shock waves. In 1897 an Englishman, John Milne, produced the first seismograph – the forerunner of today's seismometers which are used to analyse earthquakes and tremors.

Shifting continents

After charts of newly explored parts of the world first appeared it was noted that the outlines of the continents, notably the east coast of South America and the west coast of Africa, looked as though they might well have once fitted together into one vast continent.

In the early 20th century the German geophysicist Alfred Wegener attempted to explain why evidence of lush tropical prehistoric vegetation had been found in such cold regions as North America, Arctic Europe, and Antarctica. In 1910 he proposed that about 250 million years ago all the continents formed one huge mass, which he called 'Pangea'. Over millions of years these great land masses slowly shifted into their present positions. Wegener was unable to explain the cause of this shifting, but recent theories have suggested that the Earth's crust is broken up into large segments, or 'tectonic plates'. These are in constant motion, driven by the heat and pressure inside the Earth.

Alfred Wegener (1880–1930), the German geophysicist who first proposed the theory of continental drift, suggested that all the continents originated from a single huge land mass.

Below: A meteorologist reads his instruments at Halley Bay Geophysical Observatory in Antarctica. Scientists have only just begun to probe this vast ice-bound continent.

The Bricks of Life

How does each new generation inherit its parents' characteristics? Scientists trying to answer this question may have stumbled upon the key to the origin of life itself.

Children inherit half their genes from each parent. In this family one girl looks like her father, the other like her mother. So far, the baby most resembles his father.

For centuries farmers have bred plants and animals to produce varieties with the most desirable characteristics. But until the beginning of the 20th century there was no real science behind their methods. A fast racehorse would be bred to one that showed staying power; if the offspring showed both speed and staying power the happy owner would assume that the blood of the parents had mixed in suitable proportions. For there was some vague idea that inherited characteristics were carried in the blood — 'blue blood', 'bad blood', and so on were commonly referred to.

The secrets of inheritance

In 1859 Charles Darwin published his theory on the origin of animals and natural selection, but how individual animals acquired their special characteristics, and how they passed these down to their offspring, remained a mystery. 'The laws governing inheritance are quite unknown', he wrote. He came up with the theory that all the body's cells shed tiny 'gemmules' into the bloodstream, and these gemmules gathered to form the sex cells. A body cell that had been altered during life would thereafter produce similarly altered gemmules. This fitted in with the widely held belief put forward by the French naturalist Lamarck (1744–1829) that characteristics acquired during life — like huge muscles — could be passed on to offspring.

Darwin's theory had satisfactorily fitted in with the old 'blood' ideas and with Lamarck's work. But his cousin Francis Galton proved that it was incorrect by making blood transfusions between black and white rabbits. After the transfusions, if Darwin were right, the black rabbits should have had gemmules for white fur in their bloodstreams, the white rabbits gemmules for black fur. Their offspring would then be spotted. But no spotted rabbits were born. And the German August Weismann (1834–1914) struck a death-blow to Lamarckism when he cut off the tails of several generations of mice. Every succeeding generation was born with long tails — so this 'acquired characteristic' was clearly not being passed down.

This albino squirrel lacks all pigment for colouration. Its eyes appear red from tiny blood vessels, normally masked by pigmentation. Albinism is a hereditary condition, and is caused by a recessive gene; it appears only when the animal has no normal gene producing the 'plans' for pigment.

Weismann's own theory was that characteristics were passed down by a special inherited substance. But he realized that it would be quite impossible for a child to inherit every characteristic from both parents. Before the substances from both parents came together in the fertilized egg, some reduction in what is passed down had to take place. Then he read some recent research on cells. Each living cell contained thread-like structures called chromosomes — exactly the same number in almost every cell of every animal of the same species. But the female's egg and the male's sperm only had *half* the normal number of chromosomes. Could the answer be that heredity was carried by chromosomes? Weismann was rightly confident that this was so, and that each offspring inherited only half of each parent's characteristics.

Chromosomes and genes

Weismann knew nothing of the work on heredity published by Gregor Mendel in 1866. But when this was rediscovered in 1900 it fitted in with his theory. Now more and more people studied the chromosomes. Some of the most important work was done by the American T. H. Morgan. He made a detailed study of the fruit-fly *Drosophila*, which has the advantages of having eight very large, easily observed chromosomes and breeding rapidly. Morgan's experiments with generations of fruit-flies and his observations of their chromosomes proved that inherited factors were indeed determined by *genes* carried by the chromosomes. Later it was discovered that sometimes these genes could be changed or mutated by radiation from outer space. This would give rise to what farmers had long called 'sports' — creatures that for no apparent reason were different from their ancestors. All this research made it more and more clear that both Darwin and Mendel had been right.

Later research workers showed that the genes were made up of a nucleic acid, DNA, and in 1953 James Watson and Francis Crick of Cambridge presented a model showing the structure of DNA and explained how it could reproduce itself exactly. And this may have solved the problem of how life began, since DNA — the building bricks of life — could have been produced under the conditions which existed in the waters that covered the Earth many hundreds of millions of years ago.

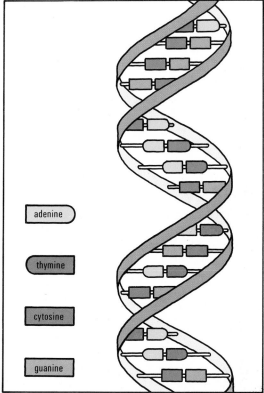

The chromosomes of a human male sorted into pairs. The 23rd 'pair' in the male consists of two very dissimilar chromosomes, an 'x' chromosome and the 'y' chromosome which determines a male.

A chromosome is made of a very long double helix (spiral) of deoxyribonucleic acid or DNA. The two strands of DNA interlock with each other; each of the four units of which the strand is made up can interlock with only one other.

adenine

thymine

cytosine

guanine

The structure of the chromosome was discovered by James Watson and Francis Crick, who won the Nobel Prize in 1962 for their work on the structure of DNA.

Index

Bold figures indicate major mention
Italic figures indicate illustration

Acknowledgements

AEI Scientific Apparatus Ltd 25 *top right*. American Museum of Natural History 26 *centre*. Heather Angel 23 *top left, top right*. Godfrey Argent/Royal Society 42 *top*. Atomic Energy Authority 5 *bottom right*, 39, 41 *top, centre, bottom*. Dr Basil Booth/ Groscience Features 44. British Antarctic Survey 45 *bottom*. British Museum 7 *top left*. Brompton Hospital 19 *bottom left*. California Institute of Technology 42 *bottom centre*. California Institute of Technology/Carnegie Institute of Washington 42 *bottom right*. Camera Press 47 *bottom*. J. Allan Cash 6 *bottom*, 8 *bottom left*. Cavendish Laboratory, Cambridge 38 *top left, bottom*. Central Electricity Generating Board 31 *bottom*. CERN 2 *top*. Gene Cox Microcolour (Hunt and Broadhurst) 33 *centre*. Department of the Environment 7 *bottom*. Mary Evans Picture Library 5 *top*, 8 *top*, 15 *top, bottom*, 16 *top right, centre*, 28 *bottom*, 29 *bottom*, 32 *top*, 33 *top*, 37 *bottom right*. Michael Holford 4 *centre*, 6 *top right*, 7 *top right*, 24 *top*. Michael Holford/National Gallery 4 *top*. Imitor 26 *top*, 45 *top right, centre*. Italian Tourist Office 14 *bottom right*. Istanbul University 11 *bottom*. Jodrell Bank 43 *top*. Lockheed Solar Observatories 42 *bottom left*. Mansell Collection 5 *centre*, 6 *top left*, 13 *top left, top right, bottom*, 15 *centre*, 16 *bottom left*, 19 *top left, top right*, 22 *top*, 26 *bottom left, bottom right*, 28 *top*, 32 *centre, bottom*, 33 *bottom*, 34 *top right*, 35 *top*, 37 *top*, 38 *top right*. Mauritshuis, The Hague 18 *top*. NASA 2 *centre*, 43 *bottom left, bottom right*, 45 *top left*. National Portrait Gallery, London 16 *bottom right*, 19 *bottom right*. Natural Science Photos 46 *bottom*. Novosti 23 *bottom*. Peabody Museum, Yale 27 *bottom*. Pope Photography Ltd 46 *top*. Radio Times Hulton Picture Library 35 *bottom*, 38 *centre*. Ronan Picture Library 8 *bottom right*, 24 *centre*. Royal Danish Embassy 40. Science Museum, London 2 *bottom*, 3 *top*, 5 *centre, bottom left*, 10 *bottom right*, 11 *top*, 14 *top left*, 16 *top left*, 18 *bottom*, 20, 24 *bottom left, bottom right*, 25 *top left, top centre*, 31 *top*, 34 *top left*, 36. Michael Short/Derby Museum and Art Gallery 12, 25 *bottom*. Tate Gallery, London 3 *bottom*, Yerkes Observatory 17.

Picture Research: Penny Warn and Jackie Newton.